LAPLACE TRANSFOR[M]

Programmes and Problems

LAPLACE TRANSFORMS

Programmes and Problems

K. A. Stroud B.Sc., Dip.Ed.

Principal Lecturer, Department of Mathematics and Statistics
The Lanchester Polytechnic, Coventry

A HALSTED PRESS BOOK

JOHN WILEY & SONS
New York

Published in the U.S.A. and Latin America by
Halsted Press, a Division of John Wiley & Sons, Inc.
New York.

ISBN: 0 470–83415–3

Library of Congress Catalog Card No.: LC 73–6317

Text set in 10/11 pt IBM Press Roman, printed by photolithography, and bound in Great Britain at The Pitman Press, Bath

PREFACE

The purpose of this book is to provide a sound introductory course in the use of Laplace transforms in the solution of differential equations and in their application to technological situations. The course requires no previous experience of the subject, but some knowledge of the solution of simple differential equations by the classical methods is assumed.

The book forms a topic module. It approaches the subject in a practical way and has been devised specifically for courses leading to

(i) B.Sc. Degree in engineering and science subjects,

(ii) Higher National Diploma and Higher National Certificate in technological subjects, and courses of a comparable standard. The module is self-contained and can therefore be introduced into any appropriate year of such courses and by its nature is equally applicable for individual or class use.

The text has been based on self-learning methods developed and extensively tested over the past ten years. Continued research into the development of efficient learning systems undertaken at the Lanchester Polytechnic, Coventry, has proved the success of the programmed approach to learning and the results obtained by student groups using these methods have been highly successful. The author's first-year undergraduate text *Engineering Mathematics* (Macmillan) based on the same techniques, has been well received, and the present work is a natural extension of these self-learning methods to a specific topic in some depth.

In controlled post-tests, each of the programmes has consistently attained a success rating in excess of 80/80, i.e. after working through each programme at least 80% of the students scored at least 80% of the possible marks. The individual nature of the method, the ability of a student to progress at his own rate, the immediate assessment of responses and, above all, the complete involvement of the student, all result in high motivation and contribute significantly to effective learning.

Programmes 1–4 establish the use of Laplace transforms in the solution of various forms of differential equations. A degree of ability with partial fractions is essential and in Programme 2 particular attention is given to an extension of the basic 'cover-up' rule to more complicated expressions. By this means, the tedious determination of constants by equating coefficients or by solving sets of simultaneous equations is avoided.

Programmes 5–8 deal with the Heaviside unit step function, periodic functions and the impulse function, while also included is an introduction to several useful theorems including the convolution theorem for the determination of inverse transforms of products. A set of worked examples provides an introduction to the application of transforms to engineering problems and the concluding section includes useful tables of transforms and inverse transforms.

Acknowledgement and thanks are due to all those who have in any way been concerned in the development and evalidation process, especially to my Head of

Department, Mr. D. A. T. Wallace, B.Sc., M.A., F.I.M.A., for his support and encouragement, and to Stanley Thornes (Publishers) Ltd. for their patient co-operation and helpful suggestions in the preparation of the text for publication.

K. A. Stroud

CONTENTS

Appendix

HINTS ON USING THE BOOK

The main part of this book consists of eight programmes covering various techniques using Laplace transforms. Each programme has been specially constructed to make learning more effective and is so designed that, provided you follow exactly the steps indicated, you are bound to learn the subject matter and the techniques involved.

The programmes are divided up into numbered sections called 'frames'. As you come to each frame, read it carefully and carry out the simple directions given. In almost every frame you will be asked to make some response, which may involve supplying part of an answer, doing one step of a problem, or working through a complete question. Whatever you are asked to do will be well within your capabilities and will build on what you have already learned. In this way, you will be carefully guided through the work as though you had a personal tutor to cover your individual needs.

When you make a response, its accuracy is immediately checked in the next frame, so that all queries are cleared up as you go along and there is no chance of your practising incorrect methods or ideas. To obtain the greatest benefit from this system, it is helpful if the following frame is covered with a sheet of paper until you have made your response.

You will soon get into it. Where a series of dots occurs, you are expected to supply the missing statement or calculation, but at all times you are free to work at your own speed. There is no need to hurry: the important thing is to be sure that you understand every step as you proceed. If you read the frames carefully and follow the directions exactly, you are bound to learn.

At the end of each programme, you will find a short Test Exercise. This covers the work of the programme in a very straightforward way and you will have no difficulty with the questions. There are no tricks and, remember, there is no need to hurry. After each test exercise, tackle as many of the Further Problems as you can – for the more practice you get the better.

A full set of answers is given at the end of the book, so you can assess and control your individual progress all the time. Many students have learned Mathematics by these methods with highly successful results. There is no reason why you should not join them.

Programme 1

TRANSFORMS OF FUNCTIONS

1

Introduction

You have no doubt already solved differential equations of the form

$$a\frac{d^2x}{dt^2}+b\frac{dx}{dt}+cx=\mathrm{F}(t),$$ where a, b, c are constant coefficients, either by the classical method of substitution and equating coefficients, or by using operator D.

In each of these methods, it is necessary first to obtain the complete general solution, i.e. both the complementary function and the particular integral, before substituting the initial conditions to establish the values of the arbitrary constants – and this can be a tedious affair.

The solution of this type of equation, particularly when the conditions at $t = 0$ are given, can be greatly speeded up by the use of LAPLACE TRANSFORMS. Furthermore, the methods that we shall be studying allow us to deal effectively with discontinuous functions, i.e. functions which suffer an abrupt change in value at some stated value of t. This is an important advantage when dealing with many practical problems.

Let us see the benefit of using Laplace transforms, by considering the solution of the equation

$$\frac{d^2x}{dt^2}+4x=t,\ \text{given that at } t=0,\ x=1 \text{ and } \frac{dx}{dt}=0.$$

We will outline the solution first by the classical method and then by using Laplace transforms, so that we can compare the two. It is highly likely that you will not understand the working in the second solution, but that, of course, is just what these programmes are all about.

So turn on to frame 2.

2

To solve $\dfrac{d^2x}{dt^2} + 4x = t$, with $x = 1$ and $\dfrac{dx}{dt} = 0$ at $t = 0$.

1. By *classical method*

(a) Complementary function:

$m^2 + 4 = 0 \quad \therefore m^2 = -4 \quad \therefore m = \pm j2$

$\therefore$ C.F. $\quad x = \text{A} \cos 2t + \text{B} \sin 2t \quad \dots$ (i)

(b) Particular integral:

Assume general form of right-hand side, i.e. $x = \text{C}t + \text{D}$

Then $\dfrac{dx}{dt} = \text{C}$ and $\dfrac{d^2x}{dt^2} = 0$.

$$4(\text{C}t + \text{D}) = t \quad \therefore \quad 4\text{C}t + 4\text{D} = t$$

Equating coefficients $4\text{C} = 1 \quad \therefore \quad \text{C} = \frac{1}{4}$ and $\text{D} = 0$

$\therefore$ P.I. is $\quad x = \frac{t}{4} \quad \dots$ (ii)

$\therefore$ General solution is $x = \text{A} \cos 2t + \text{B} \sin 2t + \dfrac{t}{4}$

Initial conditions: When $t = 0, x = 1 \quad \therefore \text{A} = 1$

Also $\quad \dfrac{dx}{dt} = -2\text{A} \sin 2t + 2\text{B} \cos 2t + \frac{1}{4}$

When $t = 0, \dfrac{dx}{dt} = 0. \quad \therefore \quad 0 = 0 + 2\text{B} + \frac{1}{4} \quad \therefore \text{B} = -\frac{1}{8}$

$\therefore$ Particular solution is $\quad \underline{x = \cos 2t - \frac{1}{8} \sin 2t + \frac{t}{4}}$

2. By *Laplace transforms*, the solution looks like this:

$$(s^2\bar{x} - s - 0) + 4\bar{x} = \frac{1}{s^2} \qquad \therefore (s^2 + 4)\bar{x} = s + \frac{1}{s^2}$$

$$\bar{x} = \frac{s}{s^2 + 4} + \frac{1}{s^2(s^2 + 4)} = \frac{s}{s^2 + 4} + \frac{1}{4}\left\{\frac{1}{s^2} - \frac{1}{s^2 + 4}\right\}$$

$$\therefore \quad \underline{x = \cos 2t - \tfrac{1}{8} \sin 2t + \tfrac{t}{4}}$$

Note: (i) The use of Laplace transforms makes the working much shorter.

(ii) This method gives the particular solution direct without having to substitute at the end to evaluate arbitrary constants.

On now to frame 3.

3

Solution of differential equations

To apply this powerful method, we work through four distinct steps:

(i) Re-write the equation in terms of its transforms,
(ii) Insert the initial conditions.
(iii) Obtain the transform of the solution by simple algebraic manipulation.
(iv) Determine the solution from its transform.

To start with, we shall deal in some detail with steps (i) and (iv) which are vitally important. Steps (ii) and (iii) are very easy and will be clear when we come to do some examples.

In this particular programme then, we are concerned with the method for expressing common functions in terms of their Laplace transforms.

So, to make a start, move on to frame 4.

4

Definition: To obtain the Laplace transform, $\mathcal{L}\{F(t)\}$, of a function F(t), we multiply the function F(t) by e^{-st} and integrate the product $e^{-st}F(t)$ with respect to t between $t = 0$ and $t = \infty$.

i.e.
$$\mathcal{L}\{F(t)\} = \int_0^\infty e^{-st}\,F(t)\,dt \qquad \ldots \quad (1)$$

The constant parameter, s, is assumed to be positive and large enough to make the product $e^{-st}F(t)$ converge to zero as $t \to \infty$. The actual value of s is not important, since this value is not directly involved in the working.

Note that (1) is a definite integral with limits to be substituted for t, so that the resulting expression will not contain t but will be expressed in terms of s only.

i.e.
$$\mathcal{L}\{F(t)\} = \int_0^\infty e^{-st}\,F(t)\,dt = f(s)$$

Make a note of this definition in your record book. You will certainly need to know it and to use it later on.

5

So, to find the Laplace transform of a function F(t) we have to multiply F(t) by e^{-st} and integrate the product with respect to t between

t = ...0...................................... and t = ...∞......................................

6

$$t = 0; \qquad t = \infty$$

$\therefore \mathcal{L}\{\mathrm{F}(t)\} = $ ....

7

$$\mathcal{L}\{\mathrm{F}(t)\} = \int_0^\infty \mathrm{e}^{-st}\mathrm{F}(t)\,dt$$

For example:

$$\mathcal{L}\{t^2\} = \int_0^\infty t^2\mathrm{e}^{-st}\,dt$$

$$\mathcal{L}\{\sin 3t\} = \int_0^\infty \sin 3t\,\mathrm{e}^{-st}\,dt$$

$\mathcal{L}\{\mathrm{e}^{5t}\} = $ ....

8

$$\mathcal{L}\{\mathrm{e}^{5t}\} = \int_0^\infty \mathrm{e}^{5t}\,\mathrm{e}^{-st}\,dt$$

So in general,

$\mathcal{L}\{\mathrm{F}(t)\} = $ ....

9

$$\mathcal{L}\{\mathrm{F}(t)\} = \int_0^\infty \mathrm{e}^{-st}\mathrm{F}(t)\,dt$$

Remember that, in this definition, s is a positive parameter, large enough to make the product $\mathrm{e}^{-st}\,\mathrm{F}(t)$ ....

....

10

. . . converge to zero, as $t \to \infty$.

Let us work through some examples.

Example 1. To find the Laplace transform of $F(t) = a$ (constant).

$$\mathcal{L}\{a\} = \int_0^\infty a\,e^{-st}\,dt = a\int_0^\infty e^{-st}\,dt$$

$$= a\left[\frac{e^{-st}}{-s}\right]_0^\infty = -\frac{a}{s}\left[e^{-st}\right]_0^\infty$$

$$= -\frac{a}{s}\left\{0 - 1\right\} = \frac{a}{s}$$

$$\therefore\ \mathcal{L}\{a\} = \frac{a}{s} \qquad \ldots \qquad (2)$$

Note that, as we said earlier, the result is a function of s (and not of t).

Example 2. To find the Laplace transform of $F(t) = e^{at}$ ('a' constant)

$$\mathcal{L}\{e^{at}\} = \int_0^\infty e^{-st}e^{at}\,dt = \int_0^\infty e^{-(s-a)t}\,dt$$

where 's' is large enough to make $e^{-(s-a)t}$ converge to zero.

$\therefore\ \mathcal{L}\{e^{at}\} =$ ....

Finish it off.

11

$$\boxed{\mathcal{L}\{e^{at}\} = \frac{1}{s-a}}$$

Here it is:

$$\mathcal{L}\{e^{at}\} = \int_0^\infty e^{-st}\, e^{at}\, dt = \int_0^\infty e^{-(s-a)t}\, dt$$

$$= \left[\frac{e^{-(s-a)t}}{-(s-a)}\right]_0^\infty = -\frac{1}{s-a}\left[e^{-(s-a)t}\right]_0^\infty$$

$$= -\frac{1}{s-a}\{0-1\} = \frac{1}{s-a}$$

$$\therefore\ \underline{\mathcal{L}\{e^{at}\} = \frac{1}{s-a}} \qquad \ldots (3)$$

So we have established two useful results:

$$\text{(i) } \mathcal{L}\{a\} = \frac{a}{s} \quad \text{and} \quad \text{(ii) } \mathcal{L}\{e^{at}\} = \frac{1}{s-a}$$

Copy these results into your record book.

12

Using the standard results you have just written down, you can now complete the following:

(i) $\mathcal{L}\{5\}$ = ...

(ii) $\mathcal{L}\{e^{3t}\}$ = ...

(iii) $\mathcal{L}\{e^{-2t}\}$ = ...

(iv) $\mathcal{L}\{1\}$ = ...

13

(i) $\mathcal{L}\{5\} = \dfrac{5}{s}$	(ii) $\mathcal{L}\{e^{3t}\} = \dfrac{1}{s-3}$
(iii) $\mathcal{L}\{e^{-2t}\} = \dfrac{1}{s+2}$	(iv) $\mathcal{L}\{1\} = \dfrac{1}{s}$

Note that the transforms of the functions are independent of 't' and are functions of 's' only.

Now another example –

Example 3. Let us find the Laplace transform of $F(t) = \cos at$.

By the definition, $\mathcal{L}\{\cos at\} = \int_0^\infty e^{-st} \cos at \, dt$ and we could determine this by using integration by parts.

However, a better way comes from remembering that

$$e^{j\theta} = \cos\theta + j\sin\theta$$

so that $\quad \cos\theta = \text{the real part of } e^{j\theta} = \mathcal{R}(e^{j\theta})$

and that $\quad \sin\theta = \text{the imaginary part of } e^{j\theta} = \mathcal{I}(e^{j\theta})$

In our present example, then, we can write

$\cos at = \ldots\ldots\ldots\ldots\ldots\ldots\ldots\ldots\ldots\ldots$

14

$$\boxed{\cos at = \mathcal{R}(e^{jat})}$$

$$\mathcal{L}\{\cos at\} = \int_0^\infty e^{-st} \cos at \, dt = \int_0^\infty e^{-st}\, \mathcal{R}(e^{jat}) \, dt$$

$$= \mathcal{R}\int_0^\infty e^{-st} e^{jat} \, dt$$

$$= \mathcal{R}\int_0^\infty e^{-(s-ja)t} \, dt$$

$$= \ldots\ldots\ldots\ldots\ldots\ldots\ldots\ldots\ldots\ldots$$

Think back to your work on complex numbers and finish it.

15

$$\boxed{\mathcal{L}\{\cos at\} = \frac{s}{s^2 + a^2}}$$

For
$$\mathcal{L}\{\cos at\} = \mathcal{R}\int_0^\infty e^{-(s-ja)t}\,dt$$

$$= \mathcal{R}\left[\frac{e^{-(s-ja)t}}{-(s-ja)}\right]_0^\infty$$

$$= \mathcal{R}\left\{-\frac{1}{(s-ja)}[0-1]\right\}$$

$$= \mathcal{R}\left\{\frac{1}{s-ja}\right\}$$

$$= \mathcal{R}\left\{\frac{s+ja}{s^2+a^2}\right\}$$

$$= \frac{s}{s^2+a^2}$$

$$\therefore\ \underline{\mathcal{L}\{\cos at\} = \frac{s}{s^2+a^2}} \qquad \ldots\ (4)$$

Note: Had we set out to find $\mathcal{L}\{\sin at\}$, then since

$$\sin at = \mathcal{I}(e^{jat})$$

the working would have been the same except that we should need the imaginary part of the result.

$$\mathcal{L}\{\sin at\} = \mathcal{I}\left\{\frac{s+ja}{s^2+a^2}\right\}$$

$$= \ldots\ldots\ldots\ldots\ldots\ldots\ldots\ldots$$

16

$$\mathcal{L}\{\sin at\} = \frac{a}{s^2 + a^2} \qquad \ldots \quad (5)$$

So far, we have established that

$$\mathcal{L}\{a\} = \frac{a}{s}; \qquad \mathcal{L}\{e^{at}\} = \frac{1}{s-a}$$

$$\mathcal{L}\{\sin at\} = \frac{a}{s^2 + a^2}; \qquad \mathcal{L}\{\cos at\} = \frac{s}{s^2 + a^2}$$

Now you can do these

(i) $\mathcal{L}\{\sin 3t\} = \ldots\ldots\ldots$ (ii) $\mathcal{L}\{\cos 2t\} = \ldots\ldots\ldots$

(iii) $\mathcal{L}\{e^{4t}\} = \ldots\ldots\ldots$ (iv) $\mathcal{L}\{7\} = \ldots\ldots\ldots$

(v) $\mathcal{L}\{e^{-t}\} = \ldots\ldots\ldots$

17

$$\text{(i)}\ \frac{3}{s^2+9}, \quad \text{(ii)}\ \frac{s}{s^2+9}, \quad \text{(iii)}\ \frac{1}{s-4}, \quad \text{(iv)}\ \frac{7}{s}, \quad \text{(v)}\ \frac{1}{s+1}$$

You must have got those right.

We have not yet considered the Laplace transforms of powers of t, so we will attend to that next.

Example 4. To find the Laplace transform of $F(t) = t$.

We have $$\mathcal{L}\{t\} = \int_0^\infty t\, e^{-st}\, dt$$

$$= \ldots\ldots\ldots\ldots \text{(integration by parts)}$$

18

$$\mathcal{L}\{t\} = \frac{1}{s^2}$$

For $\displaystyle \mathcal{L}\{t\} = \int_0^\infty t\,e^{-st}\,dt = \left[t\left(\frac{e^{-st}}{-s}\right)\right]_0^\infty + \frac{1}{s}\int_0^\infty e^{-st}\,dt$

$$= [0-0] + \frac{1}{s}\left[\frac{e^{-st}}{-s}\right]_0^\infty$$

$$= -\frac{1}{s^2}[0-1] = \frac{1}{s^2}$$

$$\therefore\ \mathcal{L}\{t\} = \frac{1}{s^2}$$

Now let us tackle $\mathcal{L}\{t^n\}$ where n is any positive integer.

19

Example 5. To determine the Laplace transform of $F(t) = t^n$ (where n is a positive integer)

We start off in the usual way:

$$\mathcal{L}\{t^n\} = \int_0^\infty t^n \, e^{-st} dt$$

$$= \left[t^n\left(\frac{e^{-st}}{-s}\right)\right]_0^\infty + \frac{n}{s}\int_0^\infty e^{-st}\, t^{n-1}\, dt$$

$$= -\frac{1}{s}\left[t^n \, e^{-st}\right]_0^\infty + \frac{n}{s}\int_0^\infty t^{n-1}\, e^{-st} dt$$

In the first term we are faced again with evaluating $t^n \, e^{-st}$ when $t \to \infty$. When $t \to \infty$, $t^n \to \infty$ while $e^{-st} \to 0$ since s is a positive integer.

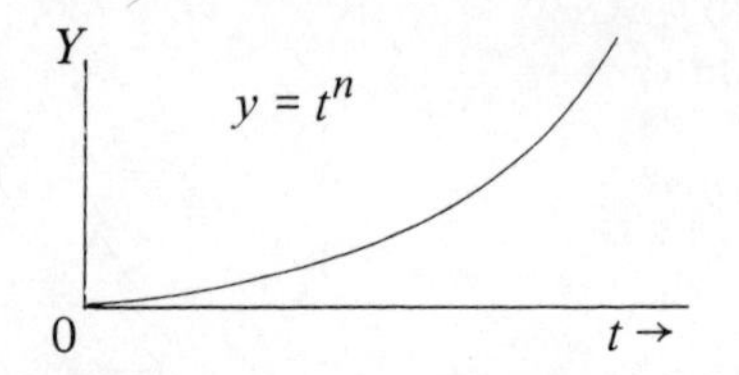

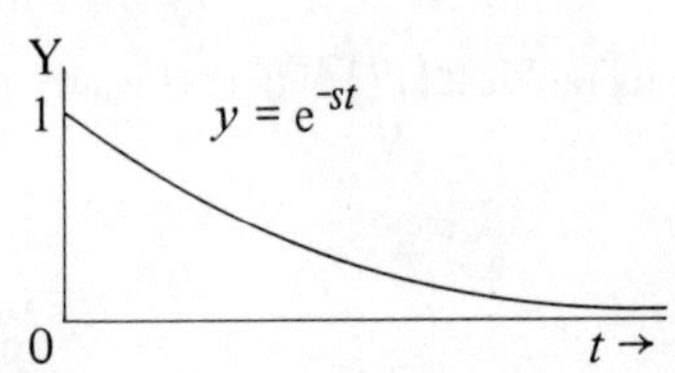

Remember we said earlier that s can be thought of as being a positive integer large enough that e^{-st} diminishes more quickly than t^n can increase, so that as $t \to \infty$, $t^n \, e^{-st} \to 0$. This we shall accept for the graph of $y = t^n \, e^{-st}$ has the general shape shown below.

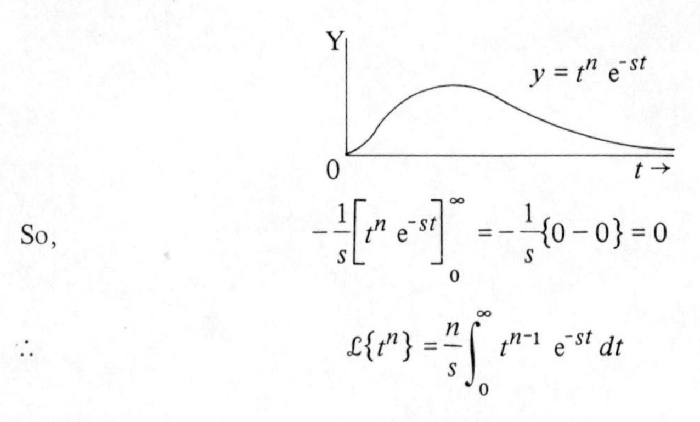

So, $$-\frac{1}{s}\left[t^n \, e^{-st}\right]_0^\infty = -\frac{1}{s}\{0 - 0\} = 0$$

$$\therefore \quad \mathcal{L}\{t^n\} = \frac{n}{s}\int_0^\infty t^{n-1}\, e^{-st} dt$$

We will finish it off in the next frame.

20

So we have

$$\mathcal{L}\{t^n\} = \int_0^\infty t^n e^{-st}\,dt = \frac{n}{s}\int_0^\infty t^{n-1} e^{-st}\,dt$$

Notice that $\int_0^\infty t^{n-1} e^{-st}\,dt$ is the same integral as $\int_0^\infty t^n e^{-st}\,dt$ except that n has been replaced by $(n-1)$.

$\therefore$ If $\quad I_n \equiv \int_0^\infty t^n e^{-st}\,dt, \quad$ then $\quad I_{n-1} \equiv \int_0^\infty t^{n-1} e^{-st}\,dt$

and our result above gives

$$I_n = \frac{n}{s}\,.\,I_{n-1} \qquad \ldots \text{(A)}$$

If we replace n by $(n-1)$ throughout this result, what do we get?

$I_{n-1} = \ldots\ldots\ldots\ldots\ldots\ldots\ldots\ldots\ldots\ldots$

21

$$\boxed{I_{n-1} = \frac{n-1}{s} I_{n-2}}$$

If we replace n by $(n-2)$ in result A we get

$$I_{n-2} = \frac{n-2}{s} I_{n-3}$$

so that

$$I_n = \frac{n}{s} I_{n-1}$$

$$= \frac{n}{s} \cdot \frac{n-1}{s} I_{n-2}$$

$$= \frac{n}{s} \cdot \frac{n-1}{s} \cdot \frac{n-2}{s} I_{n-3}$$

and so on.

Finally,

$$I_n = \frac{n}{s} \cdot \frac{n-1}{s} \cdot \frac{n-2}{s} \cdot \frac{n-3}{s} \cdot \frac{n-4}{s} \ldots \quad \ldots \quad \ldots \quad \frac{n-(n-1)}{s}$$

$$= \frac{n(n-1)(n-2)(n-3)\ldots \quad \ldots \quad \ldots 3.2.1}{s^{n+1}}$$

$$= \frac{n!}{s^{n+1}}$$

$$\therefore \mathcal{L}\{t^n\} = \frac{n!}{s^{n+1}} \qquad \ldots \quad (6)$$

Make a note of this result and then turn on.

22

$$t^n = \frac{n!}{s^{n+1}}$$

Thus $\qquad \mathcal{L}\{t^2\} = \frac{2!}{s^3} = \frac{2}{s^3} \qquad \mathcal{L}\{t^5\} = \frac{5!}{s^6} = \frac{120}{s^6}$

$\mathcal{L}\{t\} = \frac{1!}{s^2} = \frac{1}{s^2}$ (which agrees with our previous result for $\mathcal{L}\{t\}$)

....

By now, we have collected together several important standard results that we shall need to remember. Without looking at your notes, complete the following:

$\mathcal{L}\{a\}$ = ...

$\mathcal{L}\{e^{at}\}$ = ...

$\mathcal{L}\{\sin at\}$ = ...

$\mathcal{L}\{\cos at\}$ = ...

$\mathcal{L}\{t^n\}$ = ...

Then check with the next frame.

23

Here they are:

$$\mathcal{L}\{a\} = \frac{a}{s}$$

$$\mathcal{L}\{e^{at}\} = \frac{1}{s-a}$$

$$\mathcal{L}\{\sin at\} = \frac{a}{s^2+a^2}$$

$$\mathcal{L}\{\cos at\} = \frac{s}{s^2+a^2}$$

$$\mathcal{L}\{t^n\} = \frac{n!}{s^{n+1}}$$

Remember these: they will certainly be useful.

The Laplace transforms of other functions can be determined in much the same way. A more comprehensive list is given at the end of the programme.

Now on to the next stage of the work.

24

Two points worth noting:

(1) As in any integral, a constant *factor* remains unchanged in the transfer of the function.

e.g. $\mathcal{L}\{5 \sin 2t\} = 5\,\mathcal{L}\{\sin 2t\} = 5\dfrac{2}{s^2+4} = \dfrac{10}{s^2+4}.$

(2) If $\mathrm{F}(t)$ contains several terms, we simply take the transform of each term in turn.

e.g. $\mathcal{L}\{e^{2t} + \sin 3t\} = \mathcal{L}\{e^{2t}\} + \mathcal{L}\{\sin 3t\}$

$$= \frac{1}{s-2} + \frac{3}{s^2+9}$$

So what about these?

(i) $\mathcal{L}\{t^3 + e^{4t}\} =$ ....

(ii) $\mathcal{L}\{\sin 4t + \cos 4t\} =$ ....

(iii) $\mathcal{L}\{6\,e^{-3t} - 5\} =$ ....

(iv) $\mathcal{L}\{t^2 + 2t + 3\} =$ ....

25

Results:

(i) $\dfrac{3!}{s^4} + \dfrac{1}{s-4}.$

(ii) $\dfrac{4}{s^2+16} + \dfrac{s}{s^2+16} = \dfrac{s+4}{s^2+16}.$

(iii) $\dfrac{6}{s+3} - \dfrac{5}{s} = \dfrac{s-15}{s(s+3)}.$

(iv) $\dfrac{2!}{s^3} + \dfrac{2}{s^2} + \dfrac{3}{s} = \dfrac{2+2s+3s^2}{s^3}.$

If you happen to have any of them incorrect, go back and revise the appropriate part of the programme. There is no point in rushing ahead until you are quite sure of all the basic items.

If you have them all right, then move on to frame 26.

26

Example 6. To find the Laplace transforms of sinh *at* and cosh *at*.

$$\sinh at = \tfrac{1}{2}(e^{at} - e^{-at})$$
$$\cosh at = \tfrac{1}{2}(e^{at} + e^{-at})$$

Once we have these exponential definitions of sinh *at* and cosh *at*, the rest is very straight forward.

For $$\mathcal{L}\{\sinh at\} = \tfrac{1}{2}\mathcal{L}\{e^{at} - e^{-at}\}$$
$$= \text{........................}$$

Simplify the result.

$$\mathcal{L}\{\sinh at\} = \tfrac{1}{2}\mathcal{L}\{e^{at} - e^{-at}\} = \tfrac{1}{2}\left\{\frac{1}{s-a} - \frac{1}{s+a}\right\}$$
$$= \tfrac{1}{2}\left\{\frac{s+a-s+a}{(s-a)(s+a)}\right\} = \tfrac{1}{2}\left\{\frac{2a}{s^2-a^2}\right\} = \frac{a}{s^2-a^2}$$
$$\therefore \mathcal{L}\{\sinh at\} = \frac{a}{s^2-a^2}$$

27

$$\boxed{\mathcal{L}\{\sinh at\} = \frac{a}{s^2 - a^2}}$$

since $$\mathcal{L}\{\sinh at\} = \tfrac{1}{2}\mathcal{L}\{e^{at} - e^{-at}\}$$
$$= \frac{1}{2}\left\{\frac{1}{s-a} - \frac{1}{s+a}\right\}$$
$$= \frac{1}{2}\left\{\frac{s+a-s+a}{s^2-a^2}\right\}$$
$$\therefore \mathcal{L}\{\sinh at\} = \frac{a}{s^2 - a^2} \qquad \dots \quad (7)$$

Now find $\mathcal{L}\{\cosh at\}$ in the same way.

$$\mathcal{L}\{\cosh at\} = \tfrac{1}{2}\mathcal{L}\{e^{at} + e^{-at}\} = \tfrac{1}{2}\left\{\frac{1}{s-a} + \frac{1}{s+a}\right\} =$$

$$\tfrac{1}{2}\left\{\frac{s+a+s-a}{(s-a)(s+a)}\right\} = \tfrac{1}{2}\,\frac{2s}{s^2-a^2} = \frac{s}{s^2-a^2}$$
$$\therefore \mathcal{L}\{\cosh at\} = \frac{s}{s^2-a^2}$$

28

$$\mathcal{L}\{\cosh at\} = \frac{s}{s^2 - a^2}$$

Here is the working –

$$\cosh at = \tfrac{1}{2}(e^{at} + e^{-at})$$

$$\therefore \mathcal{L}\{\cosh at\} = \tfrac{1}{2}\mathcal{L}\{e^{at} + e^{-at}\}$$

$$= \frac{1}{2}\left\{\frac{1}{s-a} + \frac{1}{s+a}\right\}$$

$$= \frac{1}{2}\left\{\frac{s+a+s-a}{s^2-a^2}\right\}$$

$$\therefore \mathcal{L}\{\cosh at\} = \frac{s}{s^2 - a^2} \qquad \ldots \quad (8)$$

Make a note of results (7) and (8).

29

Theorem 1 – The First Shift Theorem

This provides a very useful rule which extends the results we have already established to a wider range of functions. It depends on the fact that the transform of $F(t)$ is a function of s, i.e. $f(s)$.

Rule:

$$\text{If } \mathcal{L}\{F(t)\} = f(s)$$

$$\text{then } \mathcal{L}\{e^{-at}\,F(t)\} = f(s+a)$$

i.e. if we know the transform of a function $F(t)$, we can write down the transform of $e^{-at}\,F(t)$ without further working, by simply writing $(s+a)$ wherever s occurs in the transform of $F(t)$.

For example: We know $\mathcal{L}\{t\} = \dfrac{1}{s^2}$

then, by the rule, $$\mathcal{L}\{e^{-4t}\,t\} = \frac{1}{(s+4)^2}$$

Again: We know $$\mathcal{L}\{\sin 3t\} = \frac{3}{s^2+9}$$

$$\therefore \mathcal{L}\{e^{-2t}\sin 3t\} = \ldots\ldots \frac{3}{(s+2)^2+9} \ldots\ldots$$

30

$$\mathcal{L}\{e^{-2t} \sin 3t\} = \frac{3}{(s+2)^2+9}$$

i.e we simply replace s by $(s + 2)$ in the transform of $\sin 3t$. Of course, the denominator could be multiplied out, giving $\dfrac{3}{s^2+4s+13}$.

Now you can do these in just the same way:

(i) $\mathcal{L}\{e^{-3t} \cos 2t\} =$ ....

(ii) $\mathcal{L}\{e^{-t}\, t^2\} =$ ....

(iii) $\mathcal{L}\{4\, e^{2t} \sin t\} =$ ....

31

This rule is important. Let us go through the results in detail.

(i) $\mathcal{L}\{e^{-3t} \cos 2t\} = ?$

We know that $\mathcal{L}\{\cos 2t\} = \dfrac{s}{s^2+4}$

$$\therefore\ \mathcal{L}\{e^{-3t} \cos 2t\} = \frac{s}{s^2+4} \text{ with } s \text{ replaced by } (s+3)$$

$$= \frac{s+3}{(s+3)^2+4} = \underline{\frac{s+3}{s^2+6s+13}}$$

(ii) $\mathcal{L}\{e^{-t}\, t^2\} = ?$

From our list of standard results $\mathcal{L}\{t^2\} = \dfrac{2!}{s^3} = \dfrac{2}{s^3}$

$$\therefore\ \mathcal{L}\{e^{-t} t^2\} = \underline{\frac{2}{(s+1)^3}}$$

(iii) $\mathcal{L}\{4e^{2t} \sin t\} = ?$

Again, we know that $\mathcal{L}\{4 \sin t\} = 4 \cdot \dfrac{1}{s^2+1} = \dfrac{4}{s^2+1}$

$$\therefore\ \mathcal{L}\{4\, e^{2t} \sin t\} = \frac{4}{(s-2)^2+1}$$

$$= \underline{\frac{4}{s^2-4s+5}}$$

32

Now by way of revision, you can easily do these. Take your time and do them carefully. No need to hurry.

Revision Exercise
Find the Laplace transforms of:

1. $\cos 3t$
2. e^{5t}
3. 4
4. $3 \sin 5t$
5. $2 + 3t^2$
6. $e^{-3t} \sin 2t$
7. $t\, e^{2t}$
8. $e^{-t} \cosh 4t$
9. $\cos 3t + 2 \sin 3t$
10. $e^{-3t}(1 + t^2)$

When you have finished them all, check with the next frame.

33

Results:

1. $\mathcal{L}\{\cos 3t\} = \dfrac{s}{s^2 + 9}.$

2. $\mathcal{L}\{e^{5t}\} = \dfrac{1}{s - 5}.$

3. $\mathcal{L}\{4\} = \dfrac{4}{s}.$

4. $\mathcal{L}\{3 \sin 5t\} = 3\dfrac{5}{s^2 + 25} = \dfrac{15}{s^2 + 25}.$

5. $\mathcal{L}\{2 + 3t^2\} = \dfrac{2}{s} + 3\,.\,\dfrac{2!}{s^3} = \dfrac{2}{s} + \dfrac{6}{s^3}.$

6. $\mathcal{L}\{e^{-3t} \sin 2t\} = \dfrac{2}{(s + 3)^2 + 4} = \dfrac{2}{s^2 + 6s + 13}.$

7. $\mathcal{L}\{t\, e^{2t}\} = \dfrac{1}{(s - 2)^2}.$

8. $\mathcal{L}\{e^{-t} \cosh 4t\} = \dfrac{s + 1}{(s + 1)^2 - 16} = \dfrac{s + 1}{s^2 + 2s - 15}.$

9. $\mathcal{L}\{\cos 3t + 2 \sin 3t\} = \dfrac{s}{s^2 + 9} + 2\dfrac{3}{s^2 + 9} = \dfrac{s + 6}{s^2 + 9}.$

10. $\mathcal{L}\{e^{-3t}(1 + t^2)\} = \dfrac{1}{s + 3} + \dfrac{2}{(s + 3)^3}.$

34

Two additional theorems

We have seen the use of the First Shift Theorem as an effective labour-saving device. There are two more theorems or rules which are worth knowing. They are useful on occasions.

Here they are stated as rules. Discussion of their proofs is set out in the appendix at the end of this book. Read through it if you are interested in the background.

If we know the Laplace transform of a function $F(t)$, then we can quickly obtain the transforms of $t \, . \, F(t)$ and $F(t)/t$.

Let us deal with the first of these in the next frame.

35

Theorem 2

If $\mathcal{L}\{F(t)\} = f(s)$, then $\mathcal{L}\{t \, . \, F(t)\} = -\dfrac{d}{ds}\{f(s)\}$.

e.g.

$$\mathcal{L}\{\sin 2t\} = \frac{2}{s^2 + 4}$$

$$\therefore \mathcal{L}\{t \, . \sin 2t\} = -\frac{d}{ds}\left\{\frac{2}{s^2+4}\right\}$$

$$= -2\left\{\frac{-2s}{(s^2+4)^2}\right\} = \frac{4s}{(s^2+4)^2}$$

e.g.

$$\mathcal{L}\{\cos at\} = \frac{s}{s^2 + a^2}$$

$$\therefore \mathcal{L}\{t \, . \cos at\} = \ldots\ldots\ldots\ldots\ldots\ldots$$

$$\mathcal{L}\{t . \cos at\} = -\frac{d}{ds}\left\{\frac{s}{s^2+a^2}\right\} = -\frac{d}{ds}\left\{s\,(s^2+a^2)^{-1}\right\} =$$

$$-\left\{s(-1)(s^2+a^2)^{-2}\,2s + (s^2+a^2)^{-1}\,1\right\} = \frac{2s^2}{(s^2+a^2)^2} - \frac{1}{(s^2+a^2)^1} =$$

$$\frac{2s^2-(s^2+a^2)}{(s^2+a^2)^2} = \frac{s^2-a^2}{(s^2+a^2)^2}$$

$$\mathcal{L}\{t . \cos at\} = \frac{s^2-a^2}{(s^2+a^2)^2}$$

36

$$\boxed{\frac{s^2 - a^2}{(s^2 + a^2)^2}}$$

For
$$\mathcal{L}\{\cos at\} = \frac{s}{s^2 + a^2}$$

$$\therefore\ \mathcal{L}\{t.\cos at\} = -\frac{d}{ds}\left\{\frac{s}{s^2 + a^2}\right\}$$
$$= -\left\{\frac{(s^2 + a^2) - s.2s}{(s^2 + a^2)^2}\right\}$$
$$= -\left\{\frac{s^2 + a^2 - 2s^2}{(s^2 + a^2)^2}\right\}$$
$$= -\left\{\frac{a^2 - s^2}{(s^2 + a^2)^2}\right\}$$
$$= \frac{s^2 - a^2}{(s^2 + a^2)^2}$$

Of course, the process can be repeated to find the transform of $t^2 \cos at$.

$$\mathcal{L}\{t^2 \cos at\} = -\frac{d}{ds}\left\{\frac{s^2 - a^2}{(s^2 + a^2)^2}\right\}$$
$$= \ldots\ldots\ldots\ldots\ldots\ldots$$

$$= -\frac{d}{ds}\left\{(s^2-a^2)(s^2+a^2)^{-2}\right\} = -\left\{(s^2-a^2)(-2)(s^2+a^2)^{-3}2s + (s^2+a^2)^{-2}2s\right\}$$

$$= \frac{(s^2-a^2)4s}{(s^2+a^2)^3} - \frac{2s}{(s^2+a^2)^2} = \frac{(s^2-a^2)4s - 2s(s^2+a^2)}{(s^2+a^2)^3}$$

$$= \frac{2s\{2(s^2-a^2) - s^2 - a^2\}}{(s^2+a^2)^3} = \frac{2s\{2s^2-2a^2-s^2-a^2\}}{(s^2+a^2)^3}$$

$$= \frac{2s(s^2-3a^2)}{(s^2+a^2)^3}$$

$$\mathcal{L}\{t^2 \cos at\} = \frac{2s(s^2-3a^2)}{(s^2+a^2)^3}$$

37

$$\mathcal{L}\{t^2 \cos at\} = \frac{2s(s^2 - 3a^2)}{(s^2 + a^2)^3}$$

Here it is:

$$\mathcal{L}\{t \cos at\} = \frac{s^2 - a^2}{(s^2 + a^2)^2}$$

$$\therefore\ \mathcal{L}\{t^2 \cos at\} = -\frac{d}{ds}\left\{\frac{s^2 - a^2}{(s^2 + a^2)^2}\right\}$$

$$= -\left\{\frac{(s^2 + a^2)^2 . 2s - (s^2 - a^2) . 2(s^2 + a^2) . 2s}{(s^2 + a^2)^4}\right\}$$

$$= -\frac{2s(s^2 + a^2)}{(s^2 + a^2)^4}\{(s^2 + a^2) - 2(s^2 - a^2)\}$$

$$= -\frac{2s}{(s^2 + a^2)^3}\{s^2 + a^2 - 2s^2 + 2a^2\}$$

$$= -\frac{2s}{(s^2 + a^2)^3}\{-s^2 + 3a^2\}$$

$$= \frac{2s(s^2 - 3a^2)}{(s^2 + a^2)^3}$$

Therefore, in general –

$$\text{If } \mathcal{L}\{F(t)\} = f(s), \quad \text{then} \quad \mathcal{L}\{t^n F(t)\} = (-1)^n . \frac{d^n}{ds^n}\{f(s)\}. \qquad \ldots \quad (9)$$

Make a note of this in your record book.

38

Here is one for you to do in like manner.

$$\mathcal{L}\{\sinh 2t\} = \frac{2}{s^2 - 4}$$

$$\therefore\ \mathcal{L}\{t \sinh 2t\} = -\frac{d}{ds}\left\{\frac{2}{s^2-4}\right\} = -2\frac{d}{ds}\{s^2-4\}^{-1} = -2\{-(s^2-4)^{-2}(2s)\} = \frac{+4s}{(s^2-4)^2}$$

$$\mathcal{L}\{t \sinh 2t\} = \frac{4s}{(s^2-4)^2}$$

39

$$\boxed{\frac{4s}{(s^2-4)^2}}$$

We have $\mathcal{L}\{\sinh 2t\} = \dfrac{2}{s^2-4}$

$$\therefore\ \mathcal{L}\{t \sinh 2t\} = -\frac{d}{ds}\left\{\frac{2}{s^2-4}\right\}$$

$$= -2\left\{-\frac{2s}{(s^2-4)^2}\right\}$$

$$\therefore\ \mathcal{L}\{t \sinh 2t\} = \frac{4s}{(s^2-4)^2}$$

Now go on one stage further and find

$\mathcal{L}\{t^2 \sinh 2t\} =$ ....

40

$$\boxed{\frac{4(3s^2-4)}{(s^2-4)^3}}$$

Here is the working:

$$\mathcal{L}\{t^2 \sinh 2t\} = -\frac{d}{ds}\left\{\frac{4s}{(s^2-4)^2}\right\}$$

$$= -\left\{\frac{(s^2-4)^2 . 4 - 4s . 2(s^2-4)\, 2s}{(s^2-4)^4}\right\}$$

$$= -\frac{4(s^2-4)}{(s^2-4)^4}\{s^2-4-4s^2\}$$

$$= -\frac{4}{(s^2-4)^3}\{-3s^2-4\}$$

$$\therefore\ \mathcal{L}\{t^2 \sinh 2t\} = \frac{4(3s^2+4)}{(s^2-4)^3}$$

Let us now attend to the second of these two additional theorems.

On then to frame 41.

Theorem 3

If $\mathcal{L}\{F(t)\} = f(s)$, then $\mathcal{L}\left\{\frac{F(t)}{t}\right\} = \int_s^\infty f(s)\, ds.$

This rule is more restricted in use, since it depends on whether the $\lim_{t\to 0}\left\{\frac{F(t)}{t}\right\}$ exists. Here is one example of its use.

Example 1. Find $\mathcal{L}\left\{\frac{\sin t}{t}\right\}$.

We know that $$\mathcal{L}\{\sin t\} = \frac{1}{s^2 + 1}$$

also that $$\lim_{t\to 0}\left\{\frac{\sin t}{t}\right\} = 1,$$

i.e. the limiting value exists.

$$\therefore\ \mathcal{L}\left\{\frac{\sin t}{t}\right\} = \int_s^\infty \frac{1}{s^2 + 1}\, ds = \left[\tan^{-1} s\right]_s^\infty$$

$$= \frac{\pi}{2} - \tan^{-1} s = \tan^{-1}\left(\frac{1}{s}\right)$$

42

Let us do another.

Example 2. Find $\mathcal{L}\left\{\dfrac{\cos 2t - \cos 3t}{t}\right\}$.

$$\mathcal{L}\{\cos 2t - \cos 3t\} = \frac{s}{s^2+4} - \frac{s}{s^2+9}$$

$$\lim_{t \to 0}\left\{\frac{\cos 2t - \cos 3t}{t}\right\} = \frac{1-1}{0} = \frac{0}{0}\text{? So apply l'Hôpital's rule.}$$

$$= \lim_{t \to 0}\left\{\frac{-2\sin 2t + 3\sin 3t}{1}\right\} = 0.$$

$$\therefore \lim_{t \to 0}\left\{\frac{F(t)}{t}\right\} \text{ exists}$$

$$\therefore \mathcal{L}\left\{\frac{\cos 2t - \cos 3t}{t}\right\} = \int_s^\infty \left(\frac{s}{(s^2+4} - \frac{s}{s^2+9}\right) ds$$

$$= \tfrac{1}{2}\ln(s^2+4) - \tfrac{1}{2}\ln(s^2+9)$$

$$= \left[\tfrac{1}{2}\ln\frac{s^2+4}{s^2+9}\right]_s^\infty$$

When $s \to \infty$, $\frac{1}{2}\ln\left(\dfrac{s^2+4}{s^2+9}\right) \to \frac{1}{2}\ln 1 = 0$

$$\therefore \mathcal{L}\left\{\frac{\cos 2t - \cos 3t}{t}\right\} = -\tfrac{1}{2}\ln\left(\frac{s^2+4}{s^2+9}\right) = \underline{\ln\sqrt{\left(\frac{s^2+9}{s^2+4}\right)}}$$

43

Here is one for you to do on your own.

Example 3. Find $\mathcal{L}\left\{\dfrac{1-\cosh t}{t}\right\}$

(i) First of all check that $\lim_{t \to 0}\left\{\dfrac{1-\cosh t}{t}\right\}$ exists.

(ii) If it does, then write down $\mathcal{L}\{1 - \cosh t\}$.

(iii) Then hence find $\mathcal{L}\left\{\dfrac{1-\cosh t}{t}\right\}$.

Off you go. When you have finished it, *turn to frame 44.*

44

$$\ln\sqrt{\left(\frac{s^2-1}{s^2}\right)}$$

Here it is set out in full:

$$\text{To find } \mathcal{L}\left\{\frac{1-\cosh t}{t}\right\}$$

(i) $\lim\limits_{t\to 0}\left\{\frac{1-\cosh t}{t}\right\} = \frac{1-1}{0} = \frac{0}{0}$? Apply l'Hôpital's rule.

$$= \lim_{t\to 0}\left\{\frac{-\sinh t}{1}\right\} = \frac{0}{1} = 0$$

$$\therefore \lim_{t\to 0}\left\{\frac{1-\cosh t}{t}\right\} \text{ exists.}$$

(ii) $\mathcal{L}\{1-\cosh t\} = \frac{1}{s} - \frac{s}{s^2-1}$

(iii) $\therefore \mathcal{L}\left\{\frac{1-\cosh t}{t}\right\} = \int_s^\infty \left(\frac{1}{s} - \frac{s}{s^2-1}\right) ds$

$$= \left[\ln s - \tfrac{1}{2}\ln(s^2-1)\right]_s^\infty$$

$$= \left[\tfrac{1}{2}\ln\frac{s^2}{s^2-1}\right]_s^\infty$$

When $s \to \infty$, then $\ln\left\{\frac{s^2}{s^2-1}\right\} \to \ln 1 = 0$

$$\therefore \mathcal{L}\left\{\frac{1-\cosh t}{t}\right\} = \tfrac{1}{2}\left\{0 - \ln\left(\frac{s^2}{s^2-1}\right)\right\} = \ln\left\{\left(\frac{s^2}{s^2-1}\right)^{-\frac{1}{2}}\right\}$$

$$= \ln\sqrt{\left(\frac{s^2-1}{s^2}\right)}$$

45

You can now deal with a variety of common functions, so here is a short exercise by way of revision. Take your time over it.

Revision Exercise

Find the Laplace transforms of the following functions:

1. $\sin 4t$
2. $\cos 3t$
3. e^{2t}
4. 3
5. $5t^3$
6. e^{-t}
7. $\sinh 2t$
8. $\cosh 5t$

That should not take long. Results are in the next frame.

46

1. $\mathcal{L}\{\sin 4t\} = \dfrac{4}{s^2 + 16}$
2. $\mathcal{L}\{\cos 3t\} = \dfrac{s}{s^2 + 9}$
3. $\mathcal{L}\{e^{2t}\} = \dfrac{1}{s - 2}$
4. $\mathcal{L}\{3\} = \dfrac{3}{s}$
5. $\mathcal{L}\{5t^3\} = \dfrac{30}{s^4}$
6. $\mathcal{L}\{e^{-t}\} = \dfrac{1}{s + 1}$
7. $\mathcal{L}\{\sinh 2t\} = \dfrac{2}{s^2 - 4}$
8. $\mathcal{L}\{\cosh 5t\} = \dfrac{s}{s^2 - 25}$

Now find the transforms of these:

9. $t^2 + 2t - 1$
10. $e^{3t}\, t^2$
11. $e^{-2t} \sin t$
12. $t \sinh 3t$
13. $\dfrac{e^{2t} - 1}{t}$
14. $e^{2t} \cos 2t$

Go carefully: then turn on for the results.

9. $\mathcal{L}\{t^2 + 2t - 1\} = \dfrac{2!}{s^3} + \dfrac{2}{s} - \dfrac{1}{s} = \dfrac{2 + 2s - s^2}{s^3}$.

10. $\mathcal{L}\{e^{3t}\, t^2\}$: $\mathcal{L}\{t^2\} = \dfrac{2!}{s^3}$ $\quad \therefore \mathcal{L}\{e^{3t}\, t^2\} = \dfrac{2!}{(s-3)^3} = \dfrac{2}{(s-3)^3}$.

11. $\mathcal{L}\{e^{-2t} \sin t\}$: $\mathcal{L}\{\sin t\} = \dfrac{1}{s^2 + 1}$

$$\therefore \mathcal{L}\{e^{-2t} \sin t\} = \frac{1}{(s+2)^2 + 1} = \frac{1}{s^2 + 4s + 5}.$$

12. $\mathcal{L}\{t \sinh 3t\}$: $\mathcal{L}\{\sinh 3t\} = \dfrac{3}{s^2 - 9}$

$$\therefore \mathcal{L}\{t \sinh 3t\} = -\frac{d}{ds}\left\{\frac{3}{s^2 - 9}\right\} = -3\left\{-\frac{2s}{(s^2 - 9)^2}\right\} = \frac{6s}{(s^2 - 9)^2}.$$

13. $\mathcal{L}\left\{\dfrac{e^{2t} - 1}{t}\right\}$: $\quad \lim\limits_{t \to 0}\left\{\dfrac{e^{2t} - 1}{t}\right\} = \dfrac{1 - 1}{0} = \dfrac{0}{0}$? Apply l'Hôpital's rule

$$= \lim_{t \to 0}\left\{\frac{2e^{2t}}{1}\right\} = 2 \qquad \therefore \text{ limit exists.}$$

$$\mathcal{L}\{e^{2t} - 1\} = \frac{1}{s-2} - \frac{1}{s} \qquad \therefore \mathcal{L}\left\{\frac{e^{2t} - 1}{t}\right\} = \int_s^\infty \left(\frac{1}{s-2} - \frac{1}{s}\right) ds$$

$$= \Big[\ln(s-2) - \ln s\Big]_s^\infty = \left[\ln\left(\frac{s-2}{s}\right)\right]_s^\infty$$

$$\lim_{s \to \infty}\left\{\ln\left(\frac{s-2}{s}\right)\right\} = \ln 1 = 0$$

$$\therefore \mathcal{L}\left\{\frac{e^{2t} - 1}{t}\right\} = \{0\} - \left\{\ln\left(\frac{s-2}{s}\right)\right\} = \ln\left(\frac{s}{s-2}\right).$$

14. $\mathcal{L}\{e^{2t} \cos 2t\}$: $\mathcal{L}\{\cos 2t\} = \dfrac{s}{s^2 + 4}$

$$\therefore \mathcal{L}\{e^{2t} \cos 2t\} = \frac{s-2}{(s-2)^2 + 4} = \frac{s-2}{s^2 - 4s + 8}.$$

48

All that now remains is the Test Exercise. Before you work through it, here is a revision summary of what we have covered in this programme. Check down it carefully and see that there is nothing that needs brushing up.

Revision Summary

1. $\mathcal{L}\{F(t)\} = \int_0^\infty e^{-st}\, F(t)\, dt.$

2. Laplace transforms of common functions.

$F(t)$	$\mathcal{L}\{F(t)\} = f(s)$
a	$\dfrac{a}{s}$
e^{at}	$\dfrac{1}{s-a}$
$\sin at$	$\dfrac{a}{s^2+a^2}$
$\cos at$	$\dfrac{s}{s^2+a^2}$
t^n	$\dfrac{n!}{s^{n+1}}$
$\sinh at$	$\dfrac{a}{s^2-a^2}$
$\cosh at$	$\dfrac{s}{s^2-a^2}$

3. *Theorem 1.* First shift theorem

$$\text{If } \mathcal{L}\{F(t)\} = f(s), \quad \text{then} \quad \mathcal{L}\{e^{-at}\, F(t)\} = f(s+a).$$

4. *Theorem 2.* Multiplying by t^n

$$\text{If } \mathcal{L}\{F(t)\} = f(s), \quad \text{then} \quad \mathcal{L}\{t^n\, F(t)\} = (-1)^n \frac{d^n}{ds^n}\{f(s)\}.$$

5. *Theorem 3.* Dividing by t

$$\text{If } \mathcal{L}\{F(t)\} = f(s) \quad \text{and} \quad \lim_{t\to 0}\left\{\frac{F(t)}{t}\right\} \text{ exists, then } \mathcal{L}\left\{\frac{F(t)}{t}\right\} = \int_s^\infty f(s)\, ds.$$

When you are ready, turn on to frame 49 and work through the Test Exercise.

Do all the questions in the exercise. They are quite straightforward and like those you have already been doing.

Work at your own speed: there is no hurry.

Test Exercise–I

Determine the Laplace transforms of the following functions:

1. 25
2. $4\,e^{3t}$
3. $\sin 5t$
4. $2\cos 3t$
5. $\cosh 2t$
6. $3\sinh t$
7. $\sin 6t + \cos 6t$
8. $2e^{-4t} - 3e^{4t}$
9. $t^4 + 3t^2 - 2$
10. $e^{-2t}\sin 3t$
11. $e^t\,t^3$
12. $e^{3t}\,(2t + t^2)$
13. $e^{-5t}\,(\cosh 2t + \sinh 2t)$
14. $t\sin 2t$
15. $t^2\sin 2t$
16. $\dfrac{1 - \cos 2t}{t}$
17. $e^{4t}\cos 2t$
18. $e^{-3t}\cosh 3t$
19. $t\cosh 2t$
20. $\dfrac{\sinh t}{t}$

50

Further Problems – I

Determine the Laplace transforms of the following functions:

1. $e^{-2t} \cos 4t$
2. $t^2 \sin 5t$
3. $e^{3t} \sin 5t$
4. $t^2 \sin 2t$
5. $t^4 e^{-2t}$
6. $\cos (wt + \theta)$
7. $\cos^2 kt$
8. $t^2 \sin nt$
9. $e^{-t} \sin (wt + \theta)$
10. $\sin wt + wt . \cos wt$
11. $1 - e^{-t} - t\,e^{-t}$
12. $\sin kt . \cos kt$
13. $t^2 \cos wt$
14. $\dfrac{t}{2a} \sin at$
15. $e^{2t}(t^2 - 5t + 6)$
16. $t \cos at$
17. $t^2 e^{3t}$
18. $e^{-2t} (3 \cos 6t - 5 \sin 6t)$
19. $\sinh 2t - \sin 2t$
20. $e^{-4t} \cosh 2t$
21. $e^{-t} \sin^2 t$
22. $\cosh^2 4t$
23. $(\sin t - \cos t)^2$
24. $t^2 \cos t$
25. $t^3 \cos t$
26. $t \sinh 2t$
27. $\dfrac{\sinh t}{t}$
28. $\dfrac{e^{-2t} - e^{-3t}}{t}$
29. $\dfrac{\cos at - \cos bt}{t}$
30. $t\,e^{-2t} \sin wt.$

Programme 2

INVERSE TRANSFORMS

1

Introduction

In the first programme of this series on Laplace transforms, we noted the importance of being able

(i) to express a function of t in terms of its transforms,
(ii) given a transform, to find the function to which it belongs.

We have dealt at some length with (i) in the previous programme. We now look at the reverse process, i.e. given a transform, to find the corresponding function of t.

For example, we know that $\mathcal{L}\{\sin at\} = \dfrac{a}{s^2 + a^2}$.

Stating this in reverse, we could say

'The function of t whose transform is $\dfrac{a}{s^2 + a^2}$ is $\sin at$.'

or, in shorter form, $\mathcal{L}^{-1}\left\{\dfrac{a}{s^2 + a^2}\right\} = \sin at$.

where $\mathcal{L}^{-1}$ denotes 'The function whose transform is . . .   . . . '

The function $\sin at$ is said to be the *inverse transform* of $\dfrac{a}{s^2 + a^2}$.

Note: The symbol $\mathcal{L}^{-1}$ is very much like $\sin^{-1}$, $\cos^{-1}$, etc. in that it does *not* indicate a reciprocal, but is a special single symbol as interpreted above.

The Inverse Transform is denoted by the symbol $\mathcal{L}^{-1}$

Since $\mathcal{L}\{e^{at}\} = \dfrac{1}{s - a}$, then $\mathcal{L}^{-1}\left\{\dfrac{1}{s - a}\right\} = e^{at}$.

So, what do these give?

(i) $\mathcal{L}^{-1}\left\{\dfrac{3}{s}\right\} =$ ....
(ii) $\mathcal{L}^{-1}\left\{\dfrac{1}{s + 4}\right\} =$ ....

(iii) $\mathcal{L}^{-1}\left\{\dfrac{5}{s^2 + 25}\right\} =$ ....
(iv) $\mathcal{L}^{-1}\left\{\dfrac{3!}{s^4} + \dfrac{s}{s^2 + 9}\right\} =$ ....

2

(i) $\mathcal{L}^{-1}\left\{\dfrac{3}{s}\right\} = 3.$	(ii) $\mathcal{L}^{-1}\left\{\dfrac{1}{s+4}\right\} = e^{-4t}.$
(iii) $\mathcal{L}^{-1}\left\{\dfrac{5}{s^2+25}\right\} = \sin 5t.$	(iv) $\mathcal{L}^{-1}\left\{\dfrac{3!}{s^4} + \dfrac{s}{s^2+9}\right\} = t^3 + \cos 3t.$

Of course, for this, we can use our list of basic transforms in reverse.

e.g. $\quad \mathcal{L}\{t^2\} = \dfrac{2}{s^3} \qquad \therefore \; \mathcal{L}^{-1}\left\{\dfrac{2}{s^3}\right\} = t^2.$

$\mathcal{L}\{e^{3t}\} = \dfrac{1}{s-3} \qquad \therefore \; \mathcal{L}^{-1}\left\{\dfrac{1}{s-3}\right\} = e^{3t}.$

$\mathcal{L}\{\sinh 2t\} = \dfrac{2}{s^2-4} \qquad \therefore \; \mathcal{L}^{-1}\left\{\dfrac{2}{s^2-4}\right\} = \sinh 2t$

$\mathcal{L}\{\cosh 3t\} = \dfrac{s}{s^2-9} \qquad \therefore$ ....

3

$$\mathcal{L}^{-1}\left\{\frac{s}{s^2-9}\right\} = \cosh 3t$$

So, with our knowledge of transforms so far, we can do two things:

(i) Given a function, we can write down the transform.
(ii) Given a transform, we can write down the function which it represents, *provided the transform can be recognized as belonging to our table.*

But what about, for example,

$$\mathcal{L}^{-1}\left\{\frac{2s-6}{(s-2)(s-4)}\right\}$$

which does not feature in the list of standard transforms?

We will soon see how to tackle this one – and others like it.

So turn on to frame 4.

4

To find $\mathcal{L}^{-1}\left\{\dfrac{2s-6}{(s-2)(s-4)}\right\}$

In such a case as this, we write the transform in simpler form by expressing it in its *partial fractions,* breaking it down into component transforms of standard types.

Thus
$$\frac{2s-6}{(s-2)(s-4)} = \frac{1}{s-2} + \frac{1}{s-4}$$

so now
$$\mathcal{L}^{-1}\left\{\frac{2s-6}{(s-2)(s-4)}\right\} = \mathcal{L}^{-1}\left\{\frac{1}{s-2}\right\} + \mathcal{L}^{-1}\left\{\frac{1}{s-4}\right\}$$
$$= \underline{e^{2t} + e^{4t}}$$

If you can find the partial fractions, the rest is easy. The whole process rests on your ability with partial fractions so we had better make sure of those first.

5

Rules of Partial Fractions

1. The numerator must be of lower degree than the denominator. If it is not, first divide out.
2. Factorize the denominator into its prime factors. The factors you get determine the shapes of the partial fractions.
3. A linear factor $(s + a)$ gives a partial fraction $\dfrac{A}{s+a}$, where A is a constant to be determined.
4. A repeated factor $(s + a)^2$ gives $\dfrac{A}{s+a} + \dfrac{B}{(s+a)^2}$.
5. Similarly $(s + a)^3$ gives $\dfrac{A}{s+a} + \dfrac{B}{(s+a)^2} + \dfrac{C}{(s+a)^3}$.
6. A quadratic factor $(s^2 + ps + q)$ gives a P.F. $\dfrac{Ps+Q}{s^2+ps+q}$.
7. $(s^2 + ps + q)^2$ gives $\dfrac{Ps+Q}{s^2+ps+q} + \dfrac{Rs+T}{(s^2+ps+q)^2}$.

So $\dfrac{s+5}{(s-3)(s+2)}$ would give partial fractions of the form

6

$$\frac{A}{s-3}+\frac{B}{s+2}$$

And $\dfrac{s^2}{(s-2)(s+3)^2}$ would give partial fractions of the form

....

Notice that there is a repeated factor in the denominator.

7

$$\frac{s^2}{(s-2)(s+3)^2} \equiv \frac{A}{s-2}+\frac{B}{s+3}+\frac{C}{(s+3)^2}$$

Now let us work through an example in full.

Example 1. To determine $\mathcal{L}^{-1}\left\{\dfrac{2s-8}{s^2-8s+15}\right\}$.

(i) Check that the numerator is of lower degree than the denominator.

(ii) That being so, we now factorize the denominator

$$\frac{2s-8}{s^2-8s+15}=\frac{2s-8}{(\quad)(\quad)}$$

What are the factors?

8

$$\frac{2s-8}{s^2-8s+15}=\frac{2s-8}{(s-3)(s-5)}$$

Therefore, the form of the partial fractions will be

....

9

$$\frac{2s-8}{(s-3)(s-5)} \equiv \frac{A}{s-3} + \frac{B}{s-5}$$

This, of course, is an identity, since the right-hand side *is* the left-hand side written in a slightly different form. Therefore the statement is true for all values of s that we choose to substitute.

First, however, to make the working easier, we multiply throughout by the denominator $(s-3)(s-5)$.

$$\therefore 2s - 8 \equiv \dots\dots\dots\dots$$

10

$$2s - 8 \equiv A(s-5) + B(s-3)$$

To find the values of the constants A and B, we now substitute a value for s that will make one of the brackets zero.

e.g. Substitute so that $(s-5) = 0$, i.e. put $s = 5$

$$\therefore 10 - 8 = A(0) + B(2) \quad \therefore 2 = 2B \quad \therefore \underline{B = 1}$$

Now substitute so that $(s-3) = 0$, i.e. put $s = 3$

$$\therefore 6 - 8 = A(-2) + B(0) \quad \therefore -2 = -2A \quad \therefore \underline{A = 1}$$

$$\therefore \frac{2s-8}{(s-3)(s-5)} \equiv \frac{1}{s-3} + \frac{1}{s-5}$$

$$\mathcal{L}^{-1}\left\{\frac{2s-8}{s^2-8s+15}\right\} = \dots\dots\dots\dots$$

11

$$\boxed{e^{3t} + e^{5t}}$$

Here is the working in full.

To find $\mathcal{L}^{-1}\left\{\dfrac{2s-8}{s^2-8s+15}\right\}$

$$\frac{2s-8}{s^2-8s+15} = \frac{2s-8}{(s-3)(s-5)} \equiv \frac{A}{s-3} + \frac{B}{s-5}$$

$$\therefore 2s - 8 \equiv A(s-5) + B(s-3)$$

Put $s = 5$ $\quad \therefore \; 2 = A(0) + B(2) \quad \therefore B = 1$

Put $s = 3$ $\quad \therefore \; -2 = A(-2) + B(0) \quad \therefore A = 1$

$$\therefore \frac{2s-8}{(s-3)(s-5)} = \frac{1}{s-3} + \frac{1}{s-5}$$

$$\mathcal{L}^{-1}\left\{\frac{2s-8}{s^2-8s+15}\right\} = \underline{e^{3t} + e^{5t}}$$

Now this one:

Example 2. Express $\dfrac{5s-8}{s(s-4)}$ in partial fractions and hence determine

$$\mathcal{L}^{-1}\left\{\frac{5s-8}{s(s-4)}\right\}.$$

(i) The numerator is of 1st degree and the denominator of the 2nd degree. Therefore, rule 1 is satisfied.

(ii) The denominator has already been factorized.

(iii) Now we have to write the forms of the partial fractions:

$$\frac{5s-8}{s(s-4)} \equiv \ldots\ldots\ldots\ldots\ldots\ldots\ldots\ldots\ldots\ldots$$

12

$$\boxed{\frac{A}{s}+\frac{B}{s-4}}$$

Now go on and evaluate the constants A and B.

13

$$\boxed{A = 2; \quad B = 3}$$

For $\qquad 5s - 8 = A(s-4) + B(s)$

Put $s = 0$ $\qquad \therefore -8 = A(-4) + B(0) \qquad \therefore A = 2$

Put $s = 4$ $\qquad \therefore\ 12 = A(0) \quad + B(4) \qquad \therefore B = 3$

$$\therefore \frac{5s-8}{s(s-4)} = \frac{2}{s} + \frac{3}{s-4}$$

$$\therefore \mathcal{L}^{-1}\left\{\frac{5s-8}{s(s-4)}\right\} = \ldots\ldots\ldots\ldots$$

Finish it off.

14

$$\boxed{\mathcal{L}^{-1}\left\{\frac{5s-8}{s(s-4)}\right\} = 2 + 3e^{4t}}$$

It is often just as easy as that.

Now here is a slightly different one.

Example 3. Determine $\mathcal{L}^{-1}\left\{\frac{s^2-2s+3}{(s-2)^3}\right\}$

Here we have the multiple factor $(s-2)^3$ in the denominator, so the form of the partial fractions will be:

$$\frac{s^2-2s+3}{(s-2)^3} \equiv \ldots\ldots\ldots\ldots$$

15

$$\frac{A}{s-2} + \frac{B}{(s-2)^2} + \frac{C}{(s-2)^3}$$

Clearing the denominator:

$$s^2 - 2s + 3 = A(s-2)^2 + B(s-2) + C$$

Put $s = 2$ $\quad \therefore 4 - 4 + 3 = A(0) + B(0) + C \qquad \therefore \underline{C = 3}$

There are no other brackets to suggest crafty substitution. When this happens, we equate coefficients of equal powers of s on each side. Start with the highest power.

$$[s^2] \qquad 1 = A \qquad \therefore \underline{A = 1}$$

Then we go to the other extreme and equate the constant terms on each side

$$[CT] \qquad 3 = 4A - 2B + C$$

$$3 = 4 \ - 2B + 3 \qquad \therefore 2B = 4 \qquad \therefore \underline{B = 2}$$

So, expressed in partial fractons,

$$\frac{s^2 - 2s + 3}{(s-2)^3} = \ldots\ldots\ldots\ldots\ldots\ldots$$

16

$$\frac{s^2 - 2s + 3}{(s-2)^3} = \frac{1}{s-2} + \frac{2}{(s-2)^2} + \frac{3}{(s-2)^3}$$

Writing the inverse transform for each term on the right-hand side, we now have

$$\mathcal{L}^{-1}\left\{\frac{s^2 - 2s + 3}{(s-2)^3}\right\} = \ldots\ldots\ldots\ldots\ldots\ldots$$

Complete it.

17

$$\mathcal{L}^{-1}\left\{\frac{s^2-2s+3}{(s-2)^3}\right\} = e^{2t} + 2e^{2t}\,t + \frac{3}{2}e^{2t}\,t^2$$
$$= e^{2t}\left\{1 + 2t + \frac{3t^2}{2}\right\}$$

For (i) $\mathcal{L}^{-1}\left\{\frac{1}{s-2}\right\} = e^{2t}$

(ii) $\mathcal{L}^{-1}\left\{\frac{1}{s^2}\right\} = t \qquad \therefore \mathcal{L}^{-1}\left\{\frac{2}{s^2}\right\} = 2t$

and since s is replaced by $(s-2)$, this indicates a factor e^{2t}. (Remember the first shift theorem from the previous programme?)

$$\mathcal{L}^{-1}\left\{\frac{2}{(s-2)^2}\right\} = \underline{2t\,e^{2t}}$$

(iii) $$\mathcal{L}^{-1}\left\{\frac{2}{s^3}\right\} = t^2 \qquad \therefore \mathcal{L}^{-1}\left\{\frac{3}{s^3}\right\} = \frac{3}{2}t^2.$$

But s is replaced by $(s-2)$

$$\mathcal{L}^{-1}\left\{\frac{3}{(s-2)^3}\right\} = \underline{\frac{3}{2}t^2\,e^{2t}}$$

$$\therefore \qquad \mathcal{L}^{-1}\left\{\frac{s^2-2s+3}{(s-2)^3}\right\} = e^{2t} + 2t\,e^{2t} + \frac{3t^2}{2}\,e^{2t}$$
$$= \underline{e^{2t}\left\{1 + 2t + \frac{3t^2}{2}\right\}}$$

Now work this one right through on your own.

Example 4. Express $\dfrac{5s^2+14s+2}{(s-2)(s+3)^2}$ in partial fractions and hence determine

$$\mathcal{L}^{-1}\left\{\frac{5s^2+14s+2}{(s-2)(s+3)^2}\right\}$$

First find the partial fractions and check with the next frame.

18

$$\boxed{\frac{2}{s-2}+\frac{3}{s+3}-\frac{1}{(s+3)^2}}$$

For $$\frac{5s^2+14s+2}{(s-2)(s+3)^2} \equiv \frac{A}{s-2}+\frac{B}{s+3}+\frac{C}{(s+3)^2}$$

$$\therefore 5s^2+14s+2 = A(s+3)^2 + B(s-2)(s+3) + C(s-2)$$

Put $s = 2$ $\qquad \therefore 50 = A(25) + B(0) + C(0) \qquad \therefore \underline{A = 2}$

Put $s = -3$ $\qquad \therefore 5 = A(0) + B(0) + C(-5) \qquad \therefore \underline{C = -1}$

Highest power $[s^2]$ $\qquad 5 = A + B \qquad \therefore 5 = 2 + B \qquad \therefore \underline{B = 3}$

$$\therefore \frac{5s^2+14s+2}{(s-2)(s+3)^2} \equiv \frac{2}{s-2}+\frac{3}{s+3}-\frac{1}{(s+3)^2}$$

So therefore $\mathcal{L}^{-1}\left\{\dfrac{5s^2+14s+2}{(s-2)(s+3)^2}\right\} =$ ....

19

$$\boxed{2\,e^{2t} + 3e^{-3t} - t\,e^{-3t}}$$

For (i) $\mathcal{L}^{-1}\left\{\dfrac{2}{s-2}\right\} = 2\,e^{2t}$

(ii) $\mathcal{L}^{-1}\left\{\dfrac{3}{s+3}\right\} = 3\,e^{-3t}$

(iii) $\mathcal{L}^{-1}\left\{\dfrac{1}{s^2}\right\} = t \qquad \therefore \mathcal{L}^{-1}\left\{\dfrac{1}{(s+3)^2}\right\} = t\,e^{-3t}$

Now here is an example with a quadratic factor.

Example 5. Express in partial fractions $\dfrac{s^2+3s-7}{(s-1)(s^2+2)}$.

Do not forget that the quadratic factor $(s^2 + 2)$ gives rise to a partial fraction of the form $\dfrac{Ps+Q}{s^2+2}$.

You do this one. Check with the next frame when you have finished.

20

$$\boxed{\frac{2s+5}{s^2+2}-\frac{1}{s-1}.}$$

Here is the working:

$$\frac{s^2+3s-7}{(s-1)(s^2+2)} \equiv \frac{A}{s-1}+\frac{Bs+C}{s^2+2}$$

$$\therefore s^2+3s-7 \equiv A(s^2+2)+(Bs+C)(s-1)$$

Put $s = 1$ $\quad\therefore -3 = A(3) + (0) \quad \therefore \underline{A = -1}$

Now equate coefficients, taking the highest and lowest powers

$[s^2]$	$1 = A + B$	$\therefore\ 1 = -1 + B$	$\therefore \underline{B = 2}$
[CT]	$-7 = 2A - C$	$\therefore -7 = -2 - C$	$\therefore \underline{C = 5}$

$$\therefore \underline{\frac{s^2+3s-7}{(s-1)(s^2+2)} = \frac{2s+5}{s^2+2}-\frac{1}{s-1}}$$

If we had now to find $\mathcal{L}^{-1}\left\{\dfrac{s^2+3s-7}{(s-1)(s^2+2)}\right\}$, we should write the right-hand side of our result above as

$$\frac{2s}{s^2+2}+\frac{5}{s^2+2}-\frac{1}{s-1}$$

so that the required inverse transform would be

....

$\mathcal{L}^{-1}\left(\frac{s}{s^2+a^2}\right) = \cos at \quad \therefore 2\mathcal{L}^{-1}\left\{\frac{s}{s^2+2^{2/2}}\right\} = 2\cos\sqrt{2}t$

$\mathcal{L}^{-1}\left(\frac{a}{s^2+a^2}\right) = \sin at \quad \therefore \frac{5}{2^{1/2}}\mathcal{L}^{-1}\left\{\frac{2^{1/2}}{s^2+2^{2/2}}\right\} = \frac{5}{\sqrt{2}}\sin\sqrt{2}t = \frac{5\sqrt{2}}{2}\sin\sqrt{2}t$

$\mathcal{L}^{-1}\left\{\frac{1}{s-a}\right\} = e^{at} \quad \therefore (-1)\mathcal{L}^{-1}\left(\frac{1}{s-1}\right) = (-1)\,e^{t}$

$\therefore 2\cos\sqrt{2}t + \frac{5\sqrt{2}}{2}\sin\sqrt{2}t - e^{t}$

OK

21

$$\boxed{2\cos\sqrt{2}t + \frac{5\sqrt{2}}{2}\sin\sqrt{2}t - e^t}$$

Since $\frac{s}{s^2+2}$ is of the form $\frac{s}{s^2+a^2}$ where $a = \sqrt{2}$

$\frac{1}{s^2+2}$ is of the form $\frac{1}{a}\cdot\frac{a}{s^2+a^2}$ where $a = \sqrt{2}$

$\frac{1}{s-1}$ is of the form $\frac{1}{s-a}$ where $a = 1$

On now to frame 22.

22

Poles and Zeros

As we have already seen in the previous examples, a Laplace transform often appears in the form $f(s) = \frac{\phi(s)}{g(s)}$, where the degree of $\phi(s)$ is less than that of $g(s)$ and where the denominator $g(s)$ is then factorized into a number of linear factors.

$$f(s) = \frac{\phi(s)}{(s-a)(s-b)(s-c)\ldots}$$

Then the values $a, b, c, \ldots$ of s that make the denominator zero and hence $f(s)$ infinite are called the *poles* of $f(s)$.

If there are no repeated factors, the poles are *simple poles.*

If any of the factors are repeated, these are then *multiple poles.*

Values of s that make the numerator $\phi(s)$ zero and hence $f(s)$ zero are called *zeros* of $f(s)$, though these do not come into our present considerations.

Examples:

$\frac{s+2}{(s-3)(s+1)}$ has simple poles at $s = 3$ and $s = -1$.

$\frac{s+5}{(s+2)^2(2s-1)}$ has a simple pole at $s = \frac{1}{2}$ and double poles at $s = -2$.

$\frac{5s-8}{s(s-4)}$ has simple poles at $s = 0$ and $s = 4$.

$\frac{s+3}{s(s+2)(s-4)(3s-1)}$ has ....

23

> Simple poles at $s = 0, s = -2, s = 4, s = \frac{1}{3}$.

The 'Cover Up' Rule

It is important to spend some time revising Partial Fractions for the whole use of Laplace transforms in the solution of differential equations depends upon a sound knowledge of these techniques.

The approach we have used so far to evaluate the constants has been a variation of the classical method. A practical way of doing the same thing without so much writing is provided by the 'cover-up' rule –which you may know already. It is particularly useful when there are simple poles – and can be applied when there is a combination of simple and multiple poles as we shall see.

Let us quote the rule by means of an example.

24

Example 1. $f(s) = \dfrac{5s-8}{s(s-4)}$ will have partial fractions $\dfrac{A}{s} + \dfrac{B}{s-4}$.

By the 'cover-up' rule, the coefficient of $\dfrac{1}{s}$, i.e. A, is given by covering up the factor s in the denominator of $f(s)$ and finding the limiting value of what remains when s (the factor covered up) $\to 0$.

$$\therefore \text{A} = \text{coeff}^{\text{t}}.\text{ of } \frac{1}{s} = \lim_{s \to 0}\left\{\frac{5s-8}{s-4}\right\} = 2.$$

Similarly, the coefficient of $\dfrac{1}{s-4}$, i.e. B, is given by covering up the factor $(s-4)$ in the denominator of $f(s)$ and finding the limiting value of what remains when $(s-4) \to 0$, i.e. when $s \to 4$.

$$\therefore \text{B} = \text{coeff}^{\text{t}}.\text{ of } \frac{1}{s-4} = \lim_{s \to 4}\left\{\frac{5s-8}{s}\right\} = \ldots\ldots\ldots\ldots$$

25

$$\boxed{B = 3}$$

$$\therefore \frac{5s-8}{s(s-4)} = \frac{2}{s} + \frac{3}{s-4}$$

Now another:

Example 2. $\dfrac{2s^2-4}{(s-3)(s-2)(s+1)} \equiv \dfrac{A}{s-3} + \dfrac{B}{s-2} + \dfrac{C}{s+1}$

$$\therefore \ A = \lim_{s\to 3}\left\{\frac{2s^2-4}{(s-2)(s+1)}\right\} = \frac{14}{(1)(4)} = \frac{7}{2}.$$

$$B = \lim_{s\to 2}\left\{\frac{2s^2-4}{(s-3)(s+1)}\right\} = \ldots\ldots\ldots\ldots$$

$$C = \ldots\ldots\ldots\ldots$$

26

$$B = \frac{4}{(-1)(3)} = -\frac{4}{3}.$$

$$C = \lim_{s\to(-1)}\left\{\frac{2s^2-4}{(s-3)(s-2)}\right\} = \frac{-2}{(-4)(-3)} = -\frac{1}{6}.$$

$$\therefore \frac{2s^2-4}{(s-3)(s-2)(s+1)} = \frac{7}{2}\cdot\frac{1}{s-3} - \frac{4}{3}\cdot\frac{1}{s-2} - \frac{1}{6}\cdot\frac{1}{s+1}$$

$$\therefore \mathcal{L}^{-1}\left\{\frac{2s^2-4}{(s-3)(s-2)(s+1)}\right\} = \ldots\ldots\ldots\ldots$$

27

$$\boxed{\frac{7}{2}e^{3t} - \frac{4}{3}e^{2t} - \frac{1}{6}e^{-t}}$$

Now here is one for you to do in the same way.

Example 3. Express $\dfrac{3s^2-7s}{(s-1)(s-2)(s-3)}$ in partial fractions.

Do that.

28

$$\frac{3}{s-3}+\frac{2}{s-2}-\frac{2}{s-1}$$

Here are the calculations:

$$\frac{3s^2-7s}{(s-1)(s-2)(s-3)} \equiv \frac{A}{s-1}+\frac{B}{s-2}+\frac{C}{s-3}$$

$$A = \lim_{s\to 1}\left\{\frac{3s^2-7s}{(s-2)(s-3)}\right\} = \frac{-4}{(-1)(-2)} = -2 \qquad \therefore A = -2$$

$$B = \lim_{s\to 2}\left\{\frac{3s^2-7s}{(s-1)(s-3)}\right\} = \frac{-2}{(1)(-1)} = 2 \qquad \therefore B = 2$$

$$C = \lim_{s\to 3}\left\{\frac{3s^2-7s}{(s-1)(s-2)}\right\} = \frac{6}{(2)(1)} = 3 \qquad \therefore C = 3$$

$$\therefore \frac{3s^2-7s}{(s-1)(s-2)(s-3)} = \frac{3}{s-3}+\frac{2}{s-2}-\frac{2}{s-1}$$

There is very little working by this method and even that can be done in one's head with a little practice.

On to frame 29.

29

Now let us look at this one.

Example 4. Express in partial fractions $\dfrac{2-11s}{(s-2)(s^2+2s+2)}$

We know that

$$\frac{2-11s}{(s-2)(s^2+2s+2)} \equiv \frac{A}{s-2}+\frac{Bs+C}{s^2+2s+2}$$

The 'cover-up' rule can still be used to find A

$$A = \lim_{s\to 2}\left\{\frac{2-11s}{s^2+2s+2}\right\} = \ldots\ldots\ldots\ldots$$

30

$$\boxed{A = \frac{-20}{10} = -2}$$

$$\therefore \frac{2 - 11s}{(s-2)(s^2 + 2s + 2)} = \frac{-2}{s-2} + \frac{Bs + C}{s^2 + 2s + 2}$$

$$\therefore 2 - 11s = (-2s^2 - 4s - 4) + (s - 2)(Bs + C)$$

Equating coefficients of highest and lowest powers,

B = ... ; C = ...

31

$$\boxed{B = 2; \qquad C = -3}$$

For $[s^2]$	$0 = -2 + B$	$\therefore B = 2$
$[CT]$	$2 = -4 - 2C$	$\therefore C = -3$

$$\therefore \frac{2 - 11s}{(s-2)(s^2 + 2s + 2)} = \ldots\ldots\ldots\ldots\ldots\ldots\ldots\ldots\ldots\ldots$$

32

$$\boxed{\frac{-2}{s-2}+\frac{2s-3}{s^2+2s+2}}$$

Do not forget the first term which you had already found.
And so on to –

Example 5. If $f(s)=\dfrac{s+3}{(s+1)^2(s-2)}$ express $f(s)$ in its partial fractions and hence find $\mathcal{L}^{-1}\{f(s)\}$.

Note the following useful dodge. Although we have a multiple pole, we can do all our work by the 'cover-up' rule, thus:

(i) $f(s)=\dfrac{s+3}{(s+1)^2(s-2)}=\dfrac{1}{s+1}\left[\dfrac{s+3}{(s+1)(s-2)}\right]$

i.e. take out the extra factor $(s+1)$ to leave a single $(s+1)$ factor in the product in the square brackets.

(ii) Next express the fraction in the square brackets in partial fractions, using the 'cover-up' rule.

$$f(s)=\frac{1}{s+1}\left[\frac{-2/3}{s+1}+\frac{5/3}{s-2}\right]$$

$$=-\frac{2}{3}\cdot\frac{1}{(s+1)^2}+\frac{5}{3}\left[\frac{1}{(s+1)(s-2)}\right]$$

(iii) Now express the new square bracket in partial fractions, again by the 'cover-up' rule.

$$f(s)=-\frac{2}{3}\cdot\frac{1}{(s+1)^2}+\frac{5}{3}\left[\frac{-1/3}{s+1}+\frac{1/3}{s-2}\right]$$

$$\therefore f(s)=-\frac{2}{3}\cdot\frac{1}{(s+1)^2}-\frac{5}{9}\cdot\frac{1}{s+1}+\frac{5}{9}\cdot\frac{1}{s-2}$$

and that part of the job is done.

Now finish off the rest of it.

$\mathcal{L}^{-1}\{f(s)\}=$ $-\frac{2}{3}te^{-t}-\frac{5}{9}e^{-t}+\frac{5}{9}e^{2t}$ ....

33

$$\frac{5}{9}\left[e^{2t} - e^{-t}\right] - \frac{2}{3}t\,e^{-t}$$

It is a very useful and quick method, since the working is always devised so that your are working with simple poles.

Let us do one more.

Example 6. Express $f(s) = \dfrac{s+1}{(s-1)^2(s-2)^2}$ in partial fractions and hence determine $\mathcal{L}^{-1}\{f(s)\}$.

As before, we first take out multiple factors, leaving a product of simple linear factors in the bracket, thus:

$$f(s) = \frac{s+1}{(s-1)^2(s-2)^2} = \frac{1}{(s-1)(s-2)}\left[\frac{s+1}{(s-1)(s-2)}\right]$$

Now put the square bracket into partial fractions.

$$f(s) = \frac{1}{(s-1)(s-2)}\left[\ldots\ldots\ldots\ldots\ldots\ldots\right]$$

34

$$f(s) = \frac{1}{(s-1)(s-2)}\left[\frac{-2}{s-1} + \frac{3}{s-2}\right]$$

Now multiply each term by $\dfrac{1}{(s-1)(s-2)}$, leaving extra factors outside in each case.

$$f(s) = \frac{1}{s-1}\left[\frac{-2}{(s-1)(s-2)}\right] + \frac{1}{s-2}\left[\frac{3}{(s-1)(s-2)}\right]$$

Now express each of these square brackets in partial fractions:

$$f(s) = \frac{1}{s-1}\left[\ldots\ldots\ldots\ldots\right] + \frac{1}{s-2}\left[\ldots\ldots\ldots\ldots\right]$$

35

$$f(s) = \frac{1}{s-1}\left[\frac{2}{s-1} + \frac{-2}{s-2}\right] + \frac{1}{s-2}\left[\frac{-3}{s-1} + \frac{3}{s-2}\right]$$

Multiply out and tidy up.

$$f(s) = \frac{2}{(s-1)^2} - \frac{2}{(s-1)(s-2)} - \frac{3}{(s-1)(s-2)} + \frac{3}{(s-2)^2}$$

$$= \frac{2}{(s-1)^2} + \frac{3}{(s-2)^2} - \frac{5}{(s-1)(s-2)}.$$

Finally re-write the last term in partial fractions.

$$f(s) = \frac{2}{(s-1)^2} + \frac{3}{(s-2)^2} - \ldots\ldots\ldots\ldots$$

36

$$5\left[\frac{-1}{s-1} + \frac{1}{s-2}\right]$$

$$\therefore f(s) = \frac{2}{(s-1)^2} + \frac{3}{(s-2)^2} + \frac{5}{s-1} - \frac{5}{s-2}$$

All that remains is to find the inverse transforms.

$$\therefore \mathcal{L}^{-1}\{f(s)\} = \ldots\ldots\ldots\ldots$$

Finish it off.

37

$$\mathcal{L}^{-1}\{f(s)\} = 5(e^t - e^{2t}) + 2t\,e^t + 3t\,e^{2t}$$

Do this one entirely on your own.

Example 7. Find $\mathcal{L}^{-1}\left\{\frac{s}{(s-1)^3(s-2)}\right\}$

Work right through it and then check with the next frame.

38

$$\mathrm{F}(t) = 2(\mathrm{e}^{2t} - \mathrm{e}^{t}) - 2t\,\mathrm{e}^{t} - \frac{t^2\,\mathrm{e}^{t}}{2}$$

Here is the working in detail:

$$f(s) = \frac{s}{(s-1)^3(s-2)}$$

$$= \frac{1}{(s-1)^2}\left[\frac{s}{(s-1)(s-2)}\right]$$

$$= \frac{1}{(s-1)^2}\left[\frac{-1}{s-1} + \frac{2}{s-2}\right]$$

$$= \frac{-1}{(s-1)^3} + \frac{1}{s-1}\left[\frac{2}{(s-1)(s-2)}\right]$$

$$= \frac{-1}{(s-1)^3} + \frac{1}{s-1}\left[\frac{-2}{s-1} + \frac{2}{s-2}\right]$$

$$= \frac{-1}{(s-1)^3} - \frac{2}{(s-1)^2} + \frac{2}{(s-1)(s-2)}$$

$$= \frac{-1}{(s-1)^3} - \frac{2}{(s-1)^2} + \frac{-2}{s-1} + \frac{2}{s-2}$$

$$\therefore f(s) = \frac{2}{s-2} - \frac{2}{s-1} - \frac{2}{(s-1)^2} - \frac{1}{(s-1)^3}$$

$$\therefore \mathrm{F}(t) = 2\,\mathrm{e}^{2t} - 2\,\mathrm{e}^{t} - 2t\,\mathrm{e}^{t} - \frac{t^2\,\mathrm{e}^{t}}{2}$$

The 'cover-up' rule can therefore be applied to a large number of cases and is always much easier and usually quicker than the more classical methods.

Let us deal with one further example; a final one and rather an important one.

So turn on to frame 39.

39

Example 8. Express $f(s) = \dfrac{s}{(s^2+1)(s^2+4)}$ in partial fractions and hence determine F(t).

$\mathcal{L}^{-1}\left\{\dfrac{1}{(s^2+1)(s^2+4)}\right\}$ would be relatively easy, for we could regard s^2 as a simple constant and express $\dfrac{1}{(s^2+1)(s^2+4)}$ in partial fractions by the cover-up rule.

$$\frac{1}{(s^2+1)(s^2+4)} \equiv \frac{A}{s^2+1} + \frac{B}{s^2+4}$$

$$A = \lim_{s^2 \to (-1)} \left\{\frac{1}{s^2+4}\right\} = \frac{1}{3}$$

$$B = \lim_{s^2 \to (-4)} \left\{\frac{1}{s^2+1}\right\} = -\frac{1}{3}$$

$$\therefore \frac{1}{(s^2+1)(s^2+4)} = \frac{1}{3}\left[\frac{1}{s^2+1} - \frac{1}{s^2+4}\right]$$

The trouble comes when we have s instead of 1 in the numerator. Have you any idea as to what we can do, – without involving j? When you have thought about it, move on to the next frame.

40

The key is simply to take the factor s out of the numerator, i.e. $s \dfrac{1}{[(s^2+1)(s^2+4)]}$, express what is inside the square brackets in partial fractions, as we did above, and then multiply the results by the factor s

$$\frac{s}{(s^2+1)(s^2+2)} = s\left[\frac{1}{(s^2+1)(s^2+4)}\right]$$

$$= s\left[\frac{1}{3}\cdot\frac{1}{s^2+1} - \frac{1}{3}\cdot\frac{1}{s^2+4}\right]$$

$$= \frac{1}{3}\cdot\frac{s}{s^2+1} - \frac{1}{3}\cdot\frac{s}{s^2+4}$$

and now we can take inverse transforms without difficulty.

So F(t) = ...

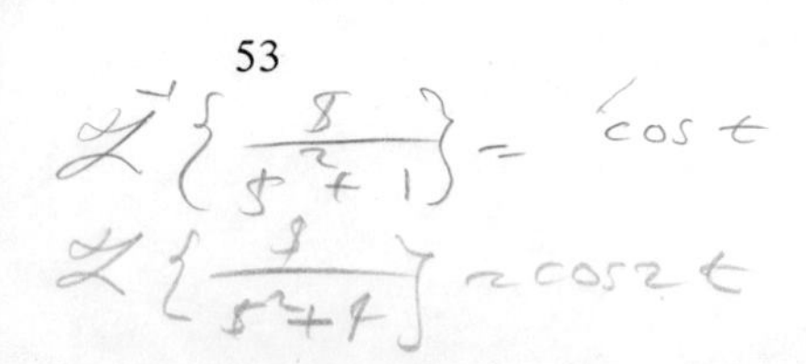

41

$$F(t) = \frac{1}{3}\cos t - \frac{1}{3}\cos 2t$$

It is a useful trick, so do this one just to be sure.

Example 9. Express $\dfrac{s}{(s^2 + 3)(s^2 + 2)}$ in partial fractions and hence determine $\mathcal{L}^{-1}\{f(s)\}$.

Work right through it and then check with frame 42

$$\mathcal{L}^{-1}\{f(s)\} = \cos(\sqrt{2}\,.\,t) - \cos(\sqrt{3}\,.\,t)$$

Here are the details:

$$f(s) = \frac{s}{(s^2 + 3)(s^2 + 2)} = s\left[\frac{1}{(s^2 + 3)(s^2 + 2)}\right]$$

$$= s\left[\frac{1/(-1)}{s^2 + 3} + \frac{1/1}{s^2 + 2}\right]$$

$$= s\left[\frac{1}{s^2 + 2} - \frac{1}{s^2 + 3}\right]$$

$$= \frac{s}{s^2 + 2} - \frac{s}{s^2 + 3}$$

$$\therefore F(t) = \mathcal{L}^{-1}\{f(s)\} = \cos(\sqrt{2}\,.\,t) - \cos(\sqrt{3}\,.\,t)$$

We have spent some time on Partial Fractions, but this merely reflects the importance of knowing how to deal with them accurately and quickly. You are now able to handle most of the functions that you are likely to meet, so let us finish this part of the programme with a few revision examples.

On to frame 43.

43

Using whichever method you like – or a mixture of them – work through the following:

Take your time and finish them all.

Revision Exercise

Express in partial fractions:

1. $\dfrac{7s-5}{(s+1)(s-2)(s-3)}$

2. $\dfrac{3s^2+8s-1}{(s-2)(s^2+s+3)}$

3. $\dfrac{2s^2+1}{s(s+1)^2}$

4. $\dfrac{s^2+5s+3}{s^2+7s+10}$

5. $\dfrac{s^2}{(s+2)^3}$

When you have finished, turn on to frame 44 to check your results.

44

Here then are the answers. See if you agree.

1. $\dfrac{4}{s-3} - \dfrac{1}{s+1} - \dfrac{3}{s-2}$.

2. $\dfrac{3}{s-2} + \dfrac{5}{s^2+s+3}$.

3. $\dfrac{1}{s} + \dfrac{1}{s+1} - \dfrac{3}{(s+1)^2}$.

4. $1 - \dfrac{1}{s+2} - \dfrac{1}{s+5}$.

5. $\dfrac{1}{s+2} - \dfrac{4}{(s+2)^2} + \dfrac{4}{(s+2)^3}$.

You must have them all, or nearly all, correct. If you have made a slip anywhere, look at your working again, see where you have gone astray and correct your mistake.

All the five questions were straightforward and covered the main types that we have dealt with.

Notice specially:

(a) The quadratic factor in Q.2, giving $\dfrac{Bs+C}{s^2+s+3}$.

(b) The zero pole and multiple poles in Q.3.

(c) The fact, that, in Q.4, the degree of the numerator is *not* less than that of the denominator. Therefore, the algebraic fraction must first be divided out.

$$\frac{s^2+5s+3}{s^2+7s+10} = 1 - \frac{2s+7}{(s+2)(s+5)}.$$

(d) The multiple pole in Q.5.

So much for partial fractions. On now to some further work on Inverse Transforms generally.

45

Inverse transforms

We can, of course, use the table of transforms established in programme 1, in reverse to give us a table of standard inverse transforms. Here they are with slight modifications, where convenient.

	$f(s)$	$F(t)$	
1.	$\dfrac{a}{s}$	a	
2.	$\dfrac{1}{s+a}$	e^{-at}	
3.	$\dfrac{n!}{s^{n+1}}$	t^n	n a positive integer
	$\dfrac{1}{s^n}$	$\dfrac{t^{n-1}}{(n-1)!}$	
4.	$\dfrac{a}{s^2+a^2}$	$\sin at$	
5.	$\dfrac{s}{s^2+a^2}$	$\cos at$	
6.	$\dfrac{a}{s^2-a^2}$	$\sinh at$	
7.	$\dfrac{s}{s^2-a^2}$	$\cosh at$	

Notice No. 3. From our table of transforms

$$\mathcal{L}\{t^n\} = \frac{n!}{s^{n+1}}$$

$$\therefore \mathcal{L}^{-1}\left\{\frac{n!}{s^{n+1}}\right\} = t^n$$

It is sometimes more useful when finding inverse transforms to re-write this result in the form

$$\mathcal{L}^{-1}\left\{\frac{1}{s^n}\right\} = \frac{t^{n-1}}{(n-1)!}$$

Both forms appear in the table above.

46

So
$$\mathcal{L}^{-1}\left\{\frac{5}{s^2+25}\right\} = \ldots\ldots\ldots\ldots$$

$$\mathcal{L}^{-1}\left\{\frac{3}{s-4}\right\} = \ldots\ldots\ldots\ldots$$

$$\mathcal{L}^{-1}\left\{\frac{2s}{s^2-9}\right\} = \ldots\ldots\ldots\ldots$$

$$\mathcal{L}^{-1}\left\{\frac{4}{s^5}\right\} = \ldots\ldots\ldots\ldots$$

$$\mathcal{L}^{-1}\left\{\frac{7}{s}\right\} = \ldots\ldots\ldots\ldots$$

$$\mathcal{L}^{-1}\left\{\frac{s}{s^2-2}\right\} = \ldots\ldots\ldots\ldots$$

47

$$\mathcal{L}^{-1}\left\{\frac{5}{s^2+25}\right\} = \sin 5t$$

$$\mathcal{L}^{-1}\left\{\frac{3}{s-4}\right\} = 3\,e^{4t}$$

$$\mathcal{L}^{-1}\left\{\frac{2s}{s^2-9}\right\} = 2\cosh 3t$$

$$\mathcal{L}^{-1}\left\{\frac{4}{s^5}\right\} = 4\,.\,\frac{t^6}{6!} = \frac{t^6}{180}$$

$$\mathcal{L}^{-1}\left\{\frac{7}{s}\right\} = 7$$

$$\mathcal{L}^{-1}\left\{\frac{s}{s^2+2}\right\} = \cos(\sqrt{2}\,.\,t)$$

We still have the *First Shift Theorem* to help us with more complicated functions of s.

Do you remember what the theorem was about?

Write it out: ....

48

> *First shift theorem*
>
> If $f(s)$ is the Laplace transform of $\text{F}(t)$ then $f(s + a)$ is the transform of $\text{e}^{-at}\,\text{F}(t)$.

This is of course, exactly the same theorem as we had in Programme 1, but, written in reverse order, it is more suitable for finding inverse transforms. *Copy it out in the form shown above.*

49

Now to apply it to some examples.

Example 1.

$$\frac{3}{s^2 + 9} \text{ is the transform of } \sin 3t$$

$$\therefore \frac{3}{(s+2)^2 + 9} \text{ is the transform of } \text{e}^{-2t} \sin 3t.$$

Example 2.

$$\mathcal{L}^{-1}\left\{\frac{s}{s^2 - 16}\right\} = \cosh 4t$$

$$\therefore \mathcal{L}^{-1}\left\{\frac{s+3}{(s+3)^2 - 16}\right\} = \ldots\ldots\ldots\ldots\ldots\ldots\ldots\ldots$$

50

> $\text{e}^{-3t} \cosh 4t$

Example 3.

$$\mathcal{L}^{-1}\left\{\frac{1}{(s+4)^5}\right\} = \ldots\ldots\ldots\ldots\ldots\ldots\ldots\ldots$$

51

$$\boxed{\frac{t^4\,e^{-4t}}{24}}$$

For
$$\mathcal{L}^{-1}\left\{\frac{1}{s^5}\right\}=\frac{t^4}{4!}$$

$$\therefore \mathcal{L}^{-1}\left\{\frac{1}{(s+4)^5}\right\}=\frac{t^4}{4!}\,.\,e^{-4t}=\underline{\frac{t^4\,e^{-4t}}{24}}$$

Now what about this one:
Example 4.

$$\mathcal{L}^{-1}\left\{\frac{1}{s^2+4s+9}\right\}$$

At first sight, this does not seem to agree with anything in our list of standard transforms. However, if we complete the square in the denominator we get

$$\mathcal{L}^{-1}\left\{\frac{1}{s^2+4s+9}\right\}=\mathcal{L}^{-1}\left\{\frac{1}{(s+2)^2+5}\right\}$$

and this is now very much like $\mathcal{L}^{-1}\left\{\frac{a}{s^2+a^2}\right\}$ with s replaced by $(s+2)$ and with $a=\sqrt{5}$.

$$\mathcal{L}^{-1}\left\{\frac{1}{(s+2)^2+5}\right\}=\ldots\ldots\ldots\ldots\ldots\ldots\ldots\ldots$$

52

$$\boxed{\frac{1}{\sqrt{5}}e^{-2t}\sin(\sqrt{5}\,.\,t)}$$

since
$$\mathcal{L}^{-1}\left\{\frac{1}{(s+2)^2+5}\right\}=\frac{1}{\sqrt{5}}\mathcal{L}^{-1}\left\{\frac{\sqrt{5}}{(s+2)^2+5}\right\}$$
$$=\underline{\frac{1}{\sqrt{5}}e^{-2t}\sin(\sqrt{5}\,.\,t)}$$

Now do this one:
Example 5.

$$\mathcal{L}^{-1}\left\{\frac{4}{s^2-6s+2}\right\}=\ldots\ldots\ldots\ldots\ldots\ldots\ldots\ldots$$

53

$$\boxed{\frac{4}{\sqrt{7}}\,e^{3t}\sinh(\sqrt{7}\,.\,t)}$$

$$\mathcal{L}^{-1}\left\{\frac{4}{s^2-6s+2}\right\}=\mathcal{L}^{-1}\left\{\frac{4}{(s-3)^2-7}\right\}$$

$$=\frac{4}{\sqrt{7}}\,\mathcal{L}^{-1}\left\{\frac{\sqrt{7}}{(s-3)^2-7}\right\}$$

$$=\frac{4}{\sqrt{7}}\,e^{3t}\ \sinh(\sqrt{7}.t)$$

And now

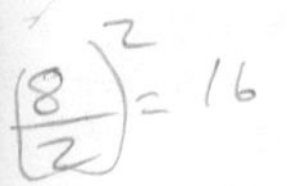

Example 6. To find $\mathcal{L}^{-1}\left\{\dfrac{s}{s^2+8s+25}\right\}$.

Completing the square in the denominator gives –

$$\mathcal{L}^{-1}\left\{\frac{s}{s^2+8s+25}\right\}=\ldots\ldots\ldots\ldots$$

54

$$\boxed{\mathcal{L}^{-1}\left\{\frac{s}{(s+4)^2+9}\right\}}$$

Now the denominator is very much like (s^2+a^2) with s replaced by $(s+4)$. The term s in the numerator must also be replaced by $(s+4)$, with an extra (-4) term introduced to correct the expression.

$$\therefore\ \mathcal{L}^{-1}\left\{\frac{s}{(s+4)^2+9}\right\}=\mathcal{L}^{-1}\left\{\frac{(s+4)-4}{(s+4)^2+9}\right\}$$

$$=\mathcal{L}^{-1}\left\{\frac{s+4}{(s+4)^2+9}\right\}-\mathcal{L}^{-1}\left\{\frac{4}{(s+4)^2+9}\right\}$$

$$=\ldots\ldots\ldots\ldots$$

Finish it off.

55

$$e^{-4t}\cos 3t - \frac{4}{3}e^{-4t}\sin 3t$$

One more:

Example 7. Determine $\mathcal{L}^{-1}\left\{\dfrac{3s}{s^2 - 2s + 26}\right\}$

Work through it completely: it is very like the last one.

56

$$3\,e^{t}\left\{\cos 5t + \frac{1}{5}\sin 5t\right\}$$

for

$$\mathcal{L}^{-1}\left\{\frac{3s}{s^2 - 2s + 26}\right\} = \mathcal{L}^{-1}\left\{\frac{3s}{(s-1)^2 + 25}\right\}$$

$$= \mathcal{L}^{-1}\left\{\frac{3(s-1) + 3}{(s-1)^2 + 5^2}\right\}$$

$$= \mathcal{L}^{-1}\left\{\frac{3(s-1)}{(s-1)^2 + 5^2}\right\} + \mathcal{L}^{-1}\left\{\frac{3}{(s-1)^2 + 5^2}\right\}$$

$$= 3\,e^{t}\cos 5t + \frac{3}{5}e^{t}\sin 5t$$

$$= 3\,e^{t}\left\{\cos 5t + \frac{1}{5}\sin 5t\right\}$$

57

And finally, this one.

Example 8. Find $\mathcal{L}^{-1}\left\{\dfrac{s-5}{s^2 + 4s + 20}\right\}$

Finish it on your own. It is another one of almost the same kind, so you will know how to tackle it.

58

$$e^{-2t}\ [\cos 4t - \tfrac{7}{4}\sin 4t]$$

Here it is in detail:

$$\mathcal{L}^{-1}\left\{\frac{s-5}{s^2+4s+20}\right\} = \mathcal{L}^{-1}\left\{\frac{s-5}{(s+2)^2+16}\right\} = \mathcal{L}^{-1}\left\{\frac{(s+2)-7}{(s+2)^2+4^2}\right\}$$

$$= \mathcal{L}^{-1}\left\{\frac{s+2}{(s+2)^2+4^2}\right\} - \mathcal{L}^{-1}\left\{\frac{7}{(s+2)^2+4^2}\right\}$$

$$= e^{-2t}\cos 4t - \tfrac{7}{4}e^{-2t}\sin 4t$$

$$= e^{-2t}\{\cos 4t - \tfrac{7}{4}\sin 4t\}$$

Do you agree? If not, check back through your working and see where you have slipped up.

Then, on to the next frame.

59

To finish up, here is a short revision exercise. Do them all.

Revision Exercise

Find the inverse transforms of

1. $\dfrac{3}{s}$
2. $\dfrac{5}{s+3}$
3. $\dfrac{4}{s^4}$
4. $\dfrac{3}{s^2+4}$
5. $\dfrac{s}{s^2+7}$
6. $\dfrac{s+2}{(s+2)^2+36}$
7. $\dfrac{5}{(s-3)^2-16}$
8. $\dfrac{3s-4}{s^2+9}$
9. $\dfrac{3s-4}{s(s-2)(s+3)}$
10. $\dfrac{s-2}{(s-1)^2(s-3)}$

When you have finished, check with the results in frame 60.

60

Here are the results. You should have had no trouble.

1. $\mathcal{L}^{-1}\left\{\frac{3}{s}\right\} = 3.$

2. $\mathcal{L}^{-1}\left\{\frac{5}{s+3}\right\} = 5\,e^{-3t}.$

3. $\mathcal{L}^{-1}\left\{\frac{4}{s^4}\right\} = 4.\frac{t^3}{3!} = \frac{2\,t^3}{3}.$

4. $\mathcal{L}^{-1}\left\{\frac{3}{s^2+4}\right\} = \frac{3}{2}\sin 2t.$

5. $\mathcal{L}^{-1}\left\{\frac{s}{s^2+7}\right\} = \cos(\sqrt{7}.t).$

6. $\mathcal{L}^{-1}\left\{\frac{s+2}{(s+2)^2+36}\right\} = e^{-2t}\cos 6t.$

7. $\mathcal{L}^{-1}\left\{\frac{5}{(s-3)^2-16}\right\} = \frac{5}{4}\,e^{3t}\sinh 4t.$

8. $\mathcal{L}^{-1}\left\{\frac{3s-4}{s^2+9}\right\} = \mathcal{L}^{-1}\left\{\frac{3s}{s^2+9}\right\} - \mathcal{L}^{-1}\left\{\frac{4}{s^2+9}\right\}$

   $= 3\cos 3t - \frac{4}{3}\sin 3t.$

9. $\mathcal{L}^{-1}\left\{\frac{3s-5}{s(s-2)(s+3)}\right\} = \frac{5}{6} + \frac{1}{10}\,e^{2t} - \frac{14}{15}\,e^{-3t}.$

10. $\mathcal{L}^{-1}\left\{\frac{s-2}{(s-1)^2(s-3)}\right\} = \frac{1}{4}\left\{e^{3t} - e^{t}\right\} + \frac{t\,e^{t}}{2}.$

That completes this programme, except for the Test Exercise which follows in frame 62. As usual, we have a Revision Summary sheet of the main points we have covered, so read down that before you go on to the test.

61 Revisionary Summary

1. *Inverse Laplace transforms* denoted by $\mathcal{L}^{-1}$.
2. *Table of inverse transforms*

$f(s)$	$F(t)$
$\dfrac{a}{s}$	a
$\dfrac{1}{s+a}$	e^{-at}
$\dfrac{1}{s^n}$	$\dfrac{t^{n-1}}{(n-1)!}$ (n a positive integer)
$\dfrac{a}{s^2+a^2}$	$\sin at$
$\dfrac{s}{s^2+a^2}$	$\cos at$
$\dfrac{a}{s^2-a^2}$	$\sinh at$
$\dfrac{s}{s^2-a^2}$	$\cosh at$

3. *Partial fractions*

(i) A linear factor $(s+a)$ gives $\dfrac{A}{s+a}$.

(ii) Repeated factors $(s+a)^2$ give $\dfrac{A}{s+a}+\dfrac{B}{(s+a)^2}$.

(iii) Similarly $(s+a)^3$ gives $\dfrac{A}{s+a}+\dfrac{B}{(s+a)^2}+\dfrac{C}{(s+a)^3}$.

(iv) Quadratic factor (s^2+ps+q) gives $\dfrac{Ps+Q}{s^2+ps+q}$.

(v) $(s^2+ps+q)^2$ gives $\dfrac{Ps+Q}{s^2+ps+q}+\dfrac{Rs+T}{(s^2+ps+q)^2}$.

4. *First shift theorem*

If $f(s)$ is the Laplace transform of $F(t)$
then $f(s+a)$ is the Laplace transform of $e^{-at}F(t)$.

62

Work steadily through the Test Exercise. Do not hurry. The problems are all very straightforward and like those you have been doing in the programme.

Test Exercise–II

1. Find the inverse transforms of

(i) $\dfrac{1}{2s-5}$ (ii) $\dfrac{3s-10}{s^2+9}$

(iii) $\dfrac{4}{(s-3)^3}$ (iv) $\dfrac{5s-4}{s^2-4}$

2. Express in partial fractions

(i) $\dfrac{13s-21}{(s-1)(s-2)(s+3)}$

(ii) $\dfrac{s+2}{s(s-3)(s^2+1)}$

(iii) $\dfrac{s+1}{(s-1)(s+2)^2}$

(iv) $\dfrac{s^2}{(s+1)^3}$

3. Find the inverse transforms of

(i) $\dfrac{s+3}{s^2+8s+16}$

(ii) $\dfrac{3s-2}{s^2-4s+20}$

(iii) $\dfrac{4s+10}{s^2-12s+32}$

(iv) $\dfrac{3s+7}{s^2-2s-3}$

63

Further Problems–II

Determine the inverse transforms of the following:

1. $\dfrac{s}{s^2+4s+8}$
2. $\dfrac{2s+1}{(s-3)^2}$
3. $\dfrac{5s-6}{(s+4)(s^2+2s+5)}$
4. $\dfrac{2s+4}{(s+3)^2}$
5. $\dfrac{s}{s^2-6s+13}$
6. $\dfrac{1}{s}\left\{\dfrac{s-a}{s+a}\right\}$
7. $\dfrac{s}{(s^2+16)^2}$
8. $\dfrac{1}{s^4-2s^3}$
9. $\dfrac{s}{(s+a)^2+b^2}$
10. $\dfrac{5s-5}{(s+3)(s^2+2s+2)}$
11. $\dfrac{s^2-10s-25}{s^3-25s}$
12. $\dfrac{3s+11}{s^2+4s+11}$
13. $\dfrac{1}{s(s^2+4s+13)}$
14. $\dfrac{s+7}{s^2+6s+10}$
15. $\dfrac{2s-1}{s^2+4s+29}$
16. $\dfrac{2s+4}{(s^2+4s+5)^2}$
17. $\dfrac{s}{s^2-2s+2}$
18. $\dfrac{3s+2}{s^2-4s-21}$
19. $\dfrac{s+1}{s^2+s+1}$
20. $\dfrac{2s-5}{s^2-9}$
21. $\dfrac{s-1}{(s+3)(s^2+2s+2)}$
22. $\dfrac{3s^2-2s-1}{(s-3)(s^2+1)}$
23. $\dfrac{3s-8}{4s^2+25}$
24. $\dfrac{s^2-3}{(s+2)(s-3)(s^2+2s+5)}$
25. $\dfrac{1}{s^2(s^2+1)(s^2+4)}$
26. $\dfrac{1}{s^2}\left\{\dfrac{s+1}{s^2+1}\right\}$
27. $\dfrac{s+3}{(s^2+6s+25)^2}$
28. $\dfrac{s^2}{(s^2+1)(s^2+2)}$
29. $\dfrac{s+2}{s^2(s+3)}$
30. $\dfrac{2s^2+s-10}{(s-4)(s^2+2s+2)}$

Programme 3

SOLUTION OF DIFFERENTIAL EQUATIONS

1

Introduction

When this series of programmes on Laplace transforms was introduced, we said that the solution of a differential equation by this method requires four distinct steps:

1. Re-write the equation in terms of its Laplace transforms.
2. Insert the given initial conditions.
3. Re-arrange the equation to give the transform of the solution.
4. Determine the inverse transform to obtain the solution.

We have concentrated so far on steps (1) and (4), and have had some real practice in writing transforms and inverse transforms. These are two vitally important processes and include a sound knowledge of partial fractions.

We are now almost ready to solve differential equations. However, if we are to express, for example, the equation

$$\frac{d^2x}{dt^2} + 3\frac{dx}{dt} + 2x = \sin 3t$$

in terms of its Laplace transforms, we must first consider the transforms of the differential coefficients, $\frac{d^2x}{dt^2}$ and $\frac{dx}{dt}$. This is the reason for the first part of the work in this programme. Then we shall be able to tackle the equations.

2

Transforms of derivatives

Let $F'(t)$ denote the first derivative of $F(t)$.

Then $\mathcal{L}\{F'(t)\} = \int_0^\infty e^{-st}\, F'(t)\, dt$ by definition.

Integrating by parts, putting $u = e^{-st}$ and $dv \equiv F'(t)\, dt$, we have

$$\mathcal{L}\{F'(t)\} = \Big[e^{-st}\, F(t)\Big]_0^\infty + s\int_0^\infty F(t)\, e^{-st}\, dt.$$

As we have so often said before:

When $t \to \infty$, then $e^{-st}\, F(t) \to$ ....

3

0

When $t \to \infty$, the product $e^{-st}\,F(t) \to 0$, since s is positive and large enough to ensure that the decay factor e^{-st} overcomes any growth characteristic of $F(t)$.

Of course, when $t = 0$, then $e^{-st}\,F(t) =$ ....

4

F(0)

For when $t = 0$, then $e^{-st}\,F(t) = e^0\ F(0) = 1.F(0) = F(0)$.

Note that the second term in the expression in frame 2, i.e. $s\int_0^\infty F(t)\,e^{-st}\,dt$, is, in fact, $s\mathcal{L}\{F(t)\}$.

$\therefore$ We have $\underline{\mathcal{L}\{F'(t)\} = -F(0) + s\mathcal{L}\{F(t)\}} = \underline{-F(0) + s\,f(s)}$

where $f(s) = \mathcal{L}\{F(t)\}$ = Laplace transform of $F(t)$,
$F(0)$ = value of $F(t)$ at $t = 0$.

So we have $$\mathcal{L}\{F'(t)\} = -F(0) + s.\mathcal{L}\{F(t)\}$$

If we now put $F'(t)$ in place of $F(t)$ in this result, we get

$$\mathcal{L}\{F''(t)\} = \text{..}$$

5

$$\boxed{\mathcal{L}\{F''(t)\} = s^2\,.\,f(s) - s\,.\,F(0) - F'(0)}$$

since
$$\begin{aligned}\mathcal{L}\{F''(t)\} &= -F'(0) + s\,.\,\mathcal{L}\{F'(t)\}\\ &= -F'(0) + s\,\{-F(0) + s\,.\,f(s)\}\\ &= -F'(0) - s\,.\,F(0) + s^2\,.\,f(s)\end{aligned}$$
$$\therefore\ \underline{\mathcal{L}\{F''(t)\} = s^2\,.\,f(s) - s\,.\,F(0) - F'(0)}$$

where, as before,
$$f(s) = \mathcal{L}\{F(t)\}$$
$$F(0) = \text{value of } F(t) \text{ at } t = 0$$
$$F'(0) = \text{value of } \frac{d}{dt}\{F(t)\} \text{ at } t = 0.$$

Turn on to frame 6.

6

So we have:

$$\mathcal{L}\{F(t)\} = f(s)$$

$$\mathcal{L}\{F'(t)\} = s.f(s) - F(0)$$

$$\mathcal{L}\{F''(t)\} = s^2.f(s) - s.F(0) - F'(0)$$

Repeating the process, we should get

$$\mathcal{L}\{F'''(t)\} = s^3.f(s) - s^2.F(0) - s.F'(0) - F''(0)$$

and you can no doubt see a clear pattern developing.

$\therefore \mathcal{L}\{F^{iv}(t)\} =$ ....

7

$$\boxed{\mathcal{L}\{F^{iv}(t)\} = s^4.f(s) - s^3.F(0) - s^2.F'(0) - s.F''(0) - F'''(0)}$$

and, in general,

$$\underline{\mathcal{L}\{F^n(t)\} = s^n.f(s) - s^{n-1}.F(0) - s^{n-2}.F'(0) - \ldots - F^{n-1}(0).}$$

So the Laplace transform of $\dfrac{d^5x}{dt^5}$ would be

$\mathcal{L}\{F^{v}(t)\} =$ ....

8

$$\boxed{\mathcal{L}\{F^{v}(t)\} = s^5.f(s) - s^4.F(0) - s^3.F'(0) - s^2.F''(0) - s.F'''(0) - F^{iv}(0).}$$

For most of the time, you will be dealing with the transforms of $\dfrac{dx}{dt}$ and $\dfrac{d^2x}{dt^2}$ – and possibly $\dfrac{d^3x}{dt^3}$ – and these are as follows;

If $\quad \mathcal{L}\{F(t)\} = f(s)$

then $\quad \mathcal{L}\{F'(t)\} = s.f(s) - F(0)$

$\mathcal{L}\{F''(t)\} =$ ....

$\mathcal{L}\{F'''(t)\} =$ ....

9

$$\mathcal{L}\{F''(t)\} = s^2.f(s) - s.F(0) - F'(0)$$
$$\mathcal{L}\{F'''(t)\} = s^3.f(s) - s^2.F(0) - s.F'(0) - F''(0)$$

Alternative notation

We have a different notation for expressing the transforms of the various differential coefficients which makes subsequent working easier to manipulate. This is the notation we shall be using from now on, so let us set it out in detail. Here it is:

$$\text{Let } x = F(t)$$

and, at $t = 0$, let $x = x_0$ i.e. $F(0) = x_0$

$$\frac{dx}{dt} = x_1 \qquad \text{i.e. } F'(0) = x_1$$

$$\frac{d^2x}{dt^2} = x_2 \qquad \text{i.e. } F''(0) = x_2$$

$$\frac{d^3x}{dt^3} = x_3 \qquad \text{i.e. } F'''(0) = x_3$$

etc.

Also, denote the Laplace transform of x by $\bar{x}$

i.e. $\bar{x} = \mathcal{L}\{x\} = \mathcal{L}\{F(t)\} = f(s)$

We can now re-write our previous results in the new notation. Using the 'dot' notation for the differential coefficients, we have

$$\mathcal{L}\{x\} = \bar{x}$$
$$\mathcal{L}\{\dot{x}\} = s.\bar{x} - x_0$$
$$\therefore \mathcal{L}\{\ddot{x}\} = \text{..................................}$$
$$\mathcal{L}\{\dddot{x}\} = \text{..................................}$$

10

Here they all are:

$$\mathcal{L}\{x\} = \bar{x}$$
$$\mathcal{L}\{\dot{x}\} = s.\bar{x} - x_0$$
$$\mathcal{L}\{\ddot{x}\} = s^2.\bar{x} - s.x_0 - x_1$$
$$\mathcal{L}\{\dddot{x}\} = s^3.\bar{x} - s^2.x_0 - s.x_1 - x_2$$

Remember that the subscript indicates the order of the differential coefficient.

e.g. x_n = the value of $\dfrac{d^n x}{dt^n}$ at $t = 0$.

Note the pattern of the results:

If we are writing the transform of the nth derivative,

(i) the first term is $s^n.\bar{x}$,
(ii) the powers of s then decrease with succeeding terms,
(iii) all terms after the first are negative,
(iv) the subscripts increase in order, x_0, x_1, x_2, etc.

So, the transform of the 4th derivative would be

$\mathcal{L}\{\ddddot{x}\}$ = ...

11

$$\mathcal{L}\{\ddddot{x}\} = s^4.\bar{x} - s^3.x_0 - s^2.x_1 - s.x_2 - x_3$$

So $\mathcal{L}\{x\} = \bar{x}$

$\mathcal{L}\{\dot{x}\}$ = ...

$\mathcal{L}\{\ddot{x}\}$ = ...

12

$$\mathcal{L}\{\dot{x}\} = s.\bar{x} - x_0$$
$$\mathcal{L}\{\ddot{x}\} = s^2.\bar{x} - s.x_0 - x_1$$

We shall be using these results quite often, so check that you have them correct and then make a note of them for future reference.

Then to frame 13.

13

Now, at last, we are ready to solve some differential equations, so let us make a start.

Solution of differential equations

Example 1. Solve $\dfrac{dx}{dt} - 4x = 8$, given that $x = 2$ at $t = 0$.

We work through the four stages that we have listed before.

(i) Re-write the whole equation in terms of Laplace transforms, remembering

that $\mathcal{L}\{x\} = \bar{x}$

$\mathcal{L}\{\dot{x}\} = \ldots\ldots\ldots\ldots$

and that $\mathcal{L}\{8\} = \ldots\ldots\ldots\ldots$

14

$$\mathcal{L}\{\dot{x}\} = s.\bar{x} - x_0$$
$$\mathcal{L}\{8\} = \frac{8}{s}$$

So the equation $\dot{x} - 4x = 8$ becomes

$\ldots\ldots\ldots\ldots$

15

$$(s.\bar{x} - x_0) - 4\bar{x} = \frac{8}{s}$$

(ii) Now insert the initial conditions.

In this example, at $t = 0$, $x = 2$, i.e. $x_0 = 2$

$$\therefore s.\bar{x} - 2 - 4\bar{x} = \frac{8}{s}$$

Collecting up terms in $\bar{x}$, gives $\quad (s-4)\bar{x} - 2 = \dfrac{8}{s}$

(iii) Now re-arrange to give an expression for $\bar{x}$

$\therefore \bar{x} = \ldots\ldots\ldots\ldots$

16

$$\bar{x} = \frac{2}{s-4} + \frac{8}{s(s-4)}$$

(iv) Finally, we take inverse transforms, by the methods we have already practised. For the first term

$$\mathcal{L}^{-1}\left\{\frac{2}{s-4}\right\} = 2\mathcal{L}^{-1}\left\{\frac{1}{s-4}\right\} = \ldots\ldots\ldots\ldots\ldots\ldots$$

17

$$2\,e^{4t}$$

The second term must be expressed in partial fractions.
Using the 'cover-up' rule,

$$\frac{8}{s(s-4)} = \frac{?}{s} + \frac{?}{s-4}$$

$$= \ldots\ldots\ldots\ldots\ldots\ldots$$

18

$$\frac{8}{s(s-4)} = \frac{-2}{s} + \frac{2}{s-4}$$

$$\therefore \mathcal{L}^{-1}\left\{\frac{8}{s(s-4)}\right\} = \ldots\ldots\ldots\ldots\ldots\ldots$$

19

$$\mathcal{L}^{-1}\left\{\frac{8}{s(s-4)}\right\} = 2\,e^{4t} - 2$$

Collecting the results together, we have $x = 2\,e^{4t} + 2\,e^{4t} - 2$

$$\therefore \quad \underline{x = 4\,e^{4t} - 2}$$

We have worked through that example step by step. Let us see it as a whole set out in the next frame.

20

Here it is again.

Solve $\dfrac{dx}{dt} - 4x = 8$, given that $x = 2$ at $t = 0$.

(i) Express in Laplace transforms

$$s.\bar{x} - x_0 - 4\bar{x} = \frac{8}{s}$$

(ii) Insert initial conditions, i.e. $x_0 = 2$

$$s.\bar{x} - 2 - 4\bar{x} = \frac{8}{s}$$

(iii) Re-arrange to give an expression for $\bar{x}$

$$(s-4)\,\bar{x} = 2 + \frac{8}{s}$$

$$\therefore \bar{x} = \frac{2}{s-4} + \frac{8}{s(s-4)}$$

(iv) Take inverse transforms

$$\bar{x} = \frac{2}{s-4} + \frac{-2}{s} + \frac{2}{s-4}$$

$$= \frac{4}{s-4} - \frac{2}{s}$$

$$\therefore \underline{x = 4\,e^{4t} - 2}$$

Now for another example

21

Example 2. Solve $2\dfrac{dx}{dt} + 3x = e^{4t}$, given that $x = 5$ when $t = 0$.

Work through the four stages.

(i) Express in Laplace transforms. The equation becomes

....

22

$$2(s\bar{x} - x_0) + 3\bar{x} = \frac{1}{s-4}$$

(ii) Now insert initial conditions, i.e. $x_0 = 5$.

$$2(s.\bar{x} - 5) + 3\bar{x} = \frac{1}{s-4}$$

Collecting up terms in $\bar{x}$

$$(2s + 3)\bar{x} - 10 = \frac{1}{s-4}$$

(iii) Re-arrange to give an expression for $\bar{x}$

$\bar{x}$ = ...

23

$$\bar{x} = \frac{10}{2s+3} + \frac{1}{(2s+3)(s-4)}$$

(iv) Finally we take inverse transforms.
The first term gives

$$\mathcal{L}^{-1}\left\{\frac{10}{2s+3}\right\} = \mathcal{L}^{-1}\left\{\frac{5}{s+3/2}\right\}$$

$=$ ....

24

$$5\,e^{-3t/2}$$

We express the second term in partial fractions, using the 'cover-up' rule.

$$\frac{1}{(2s+3)(s-4)} = \frac{?}{2s+3} + \frac{?}{s-4}$$

$=$ ....

25

$$\boxed{\frac{1}{(2s+3)(s-4)} = \frac{-2/11}{2s+3} + \frac{1/11}{s-4}}$$

$$\therefore \mathcal{L}^{-1}\left\{\frac{1}{(2s+3)(s-4)}\right\} = \mathcal{L}^{-1}\left\{\frac{1}{11}\left[\frac{1}{s-4} - \frac{2}{2s+3}\right]\right\}$$

$$= \mathcal{L}^{-1}\left\{\frac{1}{11}\left[\frac{1}{s-4} - \frac{1}{s+3/2}\right]\right\}$$

$$= \frac{1}{11}\left\{e^{4t} - e^{-3/2\,t}\right\}$$

26

$$\boxed{\frac{1}{11}\left[e^{4t} - e^{-3t/2}\right]}$$

$\therefore$ Collecting the results together

$$x = 5\,e^{-3t/2} + \frac{1}{11}\left[e^{4t} - e^{-3t/2}\right]$$

$$\therefore \quad \underline{x = \frac{1}{11}\left[54\,e^{-3t/2} + e^{4t}\right]}$$

On to the next frame.

27

Here is the complete solution.

To solve: $2\dot{x} + 3x = e^{4t}$, given that $x = 5$ when $t = 0$.

(i) $2(s.\bar{x} - x_0) + 3\bar{x} = \dfrac{1}{s-4}$

(ii) $x_0 = 5 \qquad \therefore 2(s.\bar{x} - 5) + 3\bar{x} = \dfrac{1}{s-4}$

(iii) $\therefore (2s+3)\bar{x} - 10 = \dfrac{1}{s-4}$

$$(2s+3)\bar{x} = 10 + \frac{1}{s-4}$$

$$\therefore \bar{x} = \frac{10}{2s+3} + \frac{1}{(2s+3)(s-4)}$$

$$\therefore \bar{x} = \frac{5}{s+3/2} + \frac{-2/11}{2s+3} + \frac{1/11}{s-4}$$

$$= \frac{5}{s+3/2} - \frac{1/11}{s+3/2} + \frac{1/11}{s-4}$$

$$= \frac{54}{11}\cdot\frac{1}{s+3/2} + \frac{1}{11}\cdot\frac{1}{s-4}$$

(iv) $\therefore x = \dfrac{1}{11}\left[54\,e^{-3t/2} + e^{4t}\right]$

Now for another.

28

Example 3. Solve: $3\dot{x} - 4x = \sin 2t$, given that $x = 1/3$ at $t = 0$.

(i) Expressed in transforms, the equation becomes

$3(s\bar{x} - x_0) - 4\bar{x} = \dfrac{2}{s^2+4}$ ....

29

$$3(s.\bar{x} - x_0) - 4\bar{x} = \frac{2}{s^2 + 4}$$

(ii) Insert initial conditions:

....

30

$$x_0 = 1/3 \therefore \quad 3(s.\bar{x} - 1/3) - 4\bar{x} = \frac{2}{s^2 + 4}$$

$$\therefore 3s.\bar{x} - 1 - 4\bar{x} = \frac{2}{s^2 + 4}$$

(iii) Re-arrange to give $\bar{x}$:

$\bar{x} =$ ....

31

$$(3s - 4)\bar{x} = 1 + \frac{2}{s^2 + 4}$$

$$\therefore \bar{x} = \frac{1}{3s - 4} + \frac{2}{(3s - 4)(s^2 + 4)}$$

(iv) Before we can find x, we must express the second term in partial fractions.

$$\frac{2}{(3s - 4)(s^2 + 4)}$$ ....

32

$$\frac{2}{(3s-4)(s^2+4)} \equiv \frac{18/52}{3s-4} + \frac{-6s/52 - 18/52}{s^2+4}$$
$$\equiv \frac{3}{26} \cdot \frac{1}{s-4/3} - \frac{3}{26} \cdot \frac{s}{s^2+4} - \frac{9}{26} \cdot \frac{1}{s^2+4}$$

Collecting terms together, we have:

$$\bar{x} = \frac{1}{3s-4} + \frac{3}{26} \cdot \frac{1}{s-4/3} - \frac{3}{26} \cdot \frac{s}{s^2+4} - \frac{9}{26} \cdot \frac{1}{s^2+4}$$

$$= \frac{1}{3} \cdot \frac{1}{s-4/3} + \frac{3}{26} \cdot \frac{1}{s-4/3} - \frac{3}{26} \cdot \frac{s}{s^2+4} - \frac{9}{26} \cdot \frac{1}{s^2+4}$$

$$\therefore \; \bar{x} = \frac{29}{78} \cdot \frac{1}{s-4/3} - \frac{3}{26} \left\{ \frac{s}{s^2+4} + 3 \cdot \frac{1}{s^2+4} \right\}$$

$x = \frac{29}{78} \cdot e^{4/3\,t} - \frac{3}{26} \left\{ \cos 2t + \frac{3}{2} \sin 2t \right\}$ ....

33

$$x = \frac{29}{78} e^{4t/3} - \frac{3}{26} \left[\cos 2t + \frac{3}{2} \sin 2t \right]$$

Now here is one for you to do entirely on your own. When you have completed the solution, check with the next frame. Here it is:

Example 4. Solve $\frac{dx}{dt} - 3x = t\,e^{2t}$, given that at $t = 0, x = 0$.

34

$$\boxed{x = e^{3t} - e^{2t}(1 + t)}$$

Here is the working in detail.

Solve: $\dot{x} - 3x = t \, . \, e^{2t}$, given that at $t = 0, x = 0$.

$$(s.\bar{x} - x_0) - 3\bar{x} = \frac{1}{(s-2)^2}$$

$$x_0 = 0 \qquad \therefore (s-3)\,\bar{x} = \frac{1}{(s-2)^2}$$

$$\therefore \bar{x} = \frac{1}{(s-3)(s-2)^2}$$

$$= \frac{1}{s-2}\left\{\frac{1/1}{s-3} + \frac{1/(-1)}{s-2}\right\}$$

$$= \frac{1}{(s-2)(s-3)} - \frac{1}{(s-2)^2}$$

$$= \frac{1}{s-3} - \frac{1}{s-2} - \frac{1}{(s-2)^2}$$

$$\therefore x = e^{3t} - e^{2t} - t \, . \, e^{2t}$$

$$\therefore \underline{x = e^{3t} - (1 + t)\,e^{2t}}$$

35

All first order equations are solved in very much the same way. As you see, success depends very largely on a sound knowledge of partial fractions.

Let us now apply the same general method to second order equations.

So on to frame 36.

36

Solution of second order differential equations

Example 1. Solve: $\ddot{x} - 3\dot{x} + 2x = 0$, given that at $t = 0, x = 4$ and $\dot{x} = 3$.

We work through the same four stages as before.

(i) Re-write the equation in terms of Laplace transforms, remembering that

$$\mathcal{L}\{x\} \equiv \bar{x}$$
$$\mathcal{L}\{\dot{x}\} = s.\bar{x} - x_0$$
$$\mathcal{L}\{\ddot{x}\} = s^2.\bar{x} - s.x_0 - x_1$$

So the equation in its new form becomes

$s^2\bar{x} - s.x_0 - x_1 - 3s.\bar{x} - 3x_0 + 2\bar{x} = 0$ ....

37

$$(s^2.\bar{x} - s.x_0 - x_1) - 3(s.\bar{x} - x_0) + 2\bar{x} = 0$$

(ii) Insert initial conditions.

At $t = 0, x = 4$ and $\dot{x} = 3$, i.e. $x_0 = 4, x_1 = 3$

So the equation becomes

$$(s^2.\bar{x} - 4s - 3) - 3(s.\bar{x} - 4) + 2\bar{x} = 0$$

(iii) Now re-arrange the equation by simple algebra to give

$\bar{x} =$ ....

$\bar{x}\{s^2 - 3s + 2\} = 4s + 3 - 12 = 4s - 9$

$\bar{x} = \dfrac{4s - 9}{s^2 - 3s + 2}$

38

$$\bar{x} = \frac{4s + 9}{s^2 - 3s + 2}$$

for
$$s^2.\bar{x} - 4s - 3 - 3s.\bar{x} + 12 + 2\bar{x} = 0$$
$$\therefore (s^2 - 3s + 2)\,\bar{x} - 4s + 9 = 0$$
$$(s^2 - 3s + 2)\,\bar{x} = 4s - 9$$
$$\therefore \bar{x} = \frac{4s - 9}{s^2 - 3s + 2}$$

(iv) Now to find x in terms of t we need the inverse transform of $\dfrac{4s - 9}{s^2 - 3s + 2}$ and this, of course, is where we resort to partial fractions.

$$\bar{x} = \frac{4s - 9}{(s - 1)(s - 2)} = \ldots\ldots\ldots\ldots\ldots\ldots$$

(in partial fractions)

39

$$\bar{x} = \frac{5}{s - 1} - \frac{1}{s - 2}$$

Done in your head, no doubt, by the 'cover-up' rule.

Finally
$$x = \mathcal{L}^{-1}\left\{\frac{5}{s - 1}\right\} - \mathcal{L}^{-1}\left\{\frac{1}{s - 2}\right\}$$
$$\therefore x = \ldots\ldots\ldots\ldots\ldots\ldots$$

40

$$x = 5\,e^{t} - e^{2t}$$

That was a rather simple example to start with, but all solutions go through the same four stages:

(i) Rewrite..

(ii) Insert..

(iii) Re-arrange..

(iv) Determine..

41

(i) Re-write the equation in Laplace transforms.
(ii) Insert initial conditions.
(iii) Re-arrange the equation to find $\bar{x}$.
(iv) Determine inverse transforms to find x.

Let us see the last example as a whole.

To solve: $\ddot{x} - 3\dot{x} + 2x = 0$, given that at $t = 0$, $x = 4$, $\dot{x} = 3$.

(i) $(s^2.\bar{x} - s.x_0 - x_1) - 3(s.\bar{x} - x_0) + 2\bar{x} = 0$

(ii) $x_0 = 4; \qquad x_1 = 3 \qquad \therefore (s^2.\bar{x} - 4s - 3) - 3(s.\bar{x} - 4) + 2\bar{x} = 0$

(iii) $s^2.\bar{x} - 4s - 3 - 3s.\bar{x} + 12 + 2\bar{x} = 0$

$$(s^2 - 3s + 2)\,\bar{x} - 4s + 9 = 0$$

$$\therefore \bar{x} = \frac{4s - 9}{s^2 - 3s + 2}$$

(iv) $\bar{x} = \dfrac{4s - 9}{(s-1)(s-2)} = \dfrac{5}{s-1} - \dfrac{1}{s-2}$

$$\therefore \underline{x = 5\,e^t - e^{2t}}$$

Let us do another example.

42

Example 2. Solve: $\ddot{x} + 36x = 0$, given that at $t = 0$, $x = -1$ and $\dot{x} = 2$.

We have: (i) $(s^2.\bar{x} - s.x_0 - x_1) + 36\bar{x} = 0$

(ii) $x_0 = -1; \qquad x_1 = 2.$

$$\therefore s^2.\bar{x} + s - 2 + 36\,\bar{x} = 0$$

(iii) $(s^2 + 36)\,\bar{x} + s - 2 = 0 \qquad \therefore \bar{x} = \dfrac{2 - s}{s^2 + 36}$

(iv) $\bar{x} = \dfrac{2}{s^2 + 36} - \dfrac{s}{s^2 + 36}$

$\therefore x =$ ....

43

$$x = \frac{1}{3}\sin 6t - \cos 6t$$

Well, that was easy too. Things get a little more interesting when the equation has a driving function on the right-hand side.

Example 3. Solve $\ddot{x} - 7\dot{x} + 12x = 2$, given that at $t = 0$, $x = 1$ and $\dot{x} = 5$.

We go through the normal four stages:

(i) $(s^2\bar{x} - s.x_0 - x_1) - 7(s.\bar{x} - x_0) + 12\bar{x} = \dfrac{2}{s}$

(ii) In this case $x_0 = 1$; $\quad x_1 = 5$

$$\therefore s^2.\bar{x} - s - 5 - 7s\bar{x} + 7 + 12\bar{x} = \frac{2}{s}$$

(iii) $\therefore (s^2 - 7s + 12)\bar{x} - s + 2 = \dfrac{2}{s}$

$$\therefore (s^2 - 7s + 12)\bar{x} = s - 2 + \frac{2}{s}$$

$$= \frac{s^2 - 2s + 2}{s}$$

$$\therefore \bar{x} = \frac{s^2 - 2s + 2}{s(s-3)(s-4)} = \frac{A}{s} + \frac{B}{s-3} + \frac{C}{s-4}$$

(iv) Now express in partial fractions and complete the work to find x.

Finish it off.

$A_{s\to 0} = \frac{2}{12} = \frac{1}{6}$

$B_{s\to 3} = \frac{-5}{3}$ $\qquad C_{s\to 4} = \frac{10}{4} = \frac{5}{2}$

$\bar{x} = \frac{1}{6}\left\{\frac{1}{s}\right\} - \frac{5}{3}\left\{\frac{1}{s-3}\right\} + \frac{5}{2}\left\{\frac{1}{s-4}\right\}$

$\mathcal{L}^{-1}(\bar{x}) = x = \frac{1}{6} - \frac{5}{3}e^{3t} + \frac{5}{2}e^{4t}$

44

$$\boxed{x = \frac{1}{6} - \frac{5}{3}e^{3t} + \frac{5}{2}e^{4t}}$$

For: $$\bar{x} = \frac{s^2 - 2s + 2}{s(s-3)(s-4)}$$

$$= \frac{2/12}{s} + \frac{5/(-3)}{s-3} + \frac{10/4}{s-4}$$

$$= \frac{1}{6}.\frac{1}{s} - \frac{5}{3}.\frac{1}{s-3} + \frac{5}{2}.\frac{1}{s-4}$$

$$\therefore x = \frac{1}{6} - \frac{5}{3}e^{3t} + \frac{5}{2}.e^{4t}$$

You can see how important our previous work on partial fractions was.

Another example in frame 45, so turn on.

45

Example 4. Solve: $\ddot{x} - 6\dot{x} + 8x = e^{3t}$, given that $t = 0, x = 0, \dot{x} = 2$.

(i) $(s^2.\bar{x} - s.x_0 - x_1) - 6(s.\bar{x} - x_0) + 8\bar{x} = \dfrac{1}{s-3}$

(ii) Here $x_0 = 0$; $x_1 = 2$

$$\therefore (s^2.\bar{x} - 0 - 2) - 6(s.\bar{x} - 0) + 8\bar{x} = \frac{1}{s-3}$$

$$\therefore s^2.\bar{x} - 2 - 6s.\bar{x} + 8\bar{x} = \frac{1}{s-3}$$

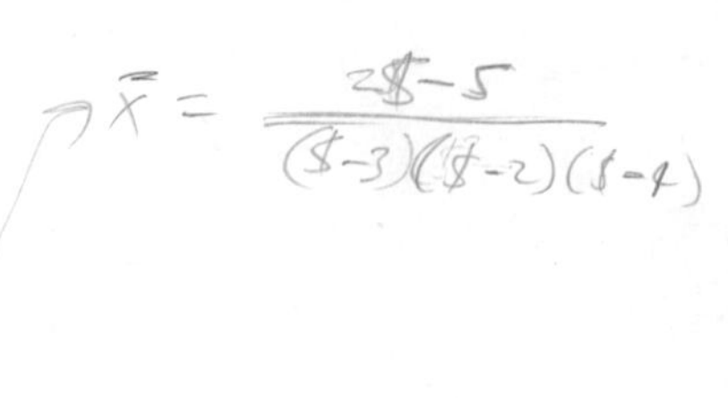

$$(s^2 - 6s + 8)\bar{x} = 2 + \frac{1}{s-3}$$

$$= \frac{2s-5}{s-3}$$

Now you complete the rest, so that finally

$x = $ ....

46

$$\boxed{x = \frac{3}{2}e^{4t} - e^{3t} - \frac{1}{2}e^{2t}}$$

We had

$$(s^2 - 6s + 8)\,\bar{x} = \frac{2s-5}{s-3}$$

$$\therefore (s-2)(s-4)\bar{x} = \frac{2s-5}{s-3}$$

$$\therefore \bar{x} = \frac{2s-5}{(s-2)(s-3)(s-4)}$$

$$\therefore \bar{x} = \frac{-1/2}{s-2} + \frac{1/(-1)}{s-3} + \frac{3/2}{s-4}$$

by the 'cover-up' method.

$$\therefore x = \frac{3}{2}e^{4t} - e^{3t} - \frac{1}{2}e^{2t}$$

Now on to frame 47 for Example 5.

47

Example 5. Solve: $\ddot{x} + x = 6 \cos 2t$, given that at $t = 0, x = 3$ and $\dot{x} = 1$.

(i) First re-write in terms of Laplace transforms (Stage 1).

What do you get?

48

$$\boxed{(s^2.\bar{x} - sx_0 - x_1) + \bar{x} = \frac{6s}{s^2+4}}$$

(ii) Now we have to insert initial conditions. (Stage 2.)

In this case x_0 = ...3....; x_1 = ...1....

49

$$\boxed{x_0 = 3; \qquad x_1 = 1}$$

Substitute these values and tidy up the left-hand side, giving

....$\bar{x}(s^2+1) - 3s - 1 = 6\frac{s}{s^2+4}$....

50

$$\boxed{(s^2 + 1)\bar{x} - 3s - 1 = \frac{6s}{s^2 + 4}}$$

since

$$(s^2.\bar{x} - 3s - 1) + \bar{x} = \frac{6s}{s^2 + 4}$$

$$(s^2 + 1)\bar{x} - 3s - 1 = \frac{6s}{s^2 + 4}$$

$\therefore (s^2 + 1)\bar{x} =$ ....

51

$$\boxed{(s^2 + 1)\bar{x} = 3s + 1 + \frac{6s}{s^2 + 4}}$$

At this point, we often have to decide whether or not to combine the right-hand side into a single algebraic fraction, or to leave it as separate terms.

In this example, we have, of course, to divide through by $(s^2 + 1)$ to obtain an expression for $\bar{x}$ and we notice that $\dfrac{3s}{s^2 + 1}$ and $\dfrac{1}{s^2 + 1}$ are closely related to standard transforms. In this case, it would be wasted effort to combine the terms, only to have to break the result down into partial fractions immediately afterwards.

$$\therefore \bar{x} = \frac{3s}{s^2 + 1} + \frac{1}{s^2 + 1} + \frac{6s}{(s^2 + 1)(s^2 + 4)}$$

We leave the first two terms as they are, but what about the third term? We dealt with this form as a special type in Programme 2 of the series. Remember?

$\dfrac{6s}{(s^2 + 1)(s^2 + 4)}$ is written as ....

52

$$\boxed{\frac{6s}{(s^2 + 1)(s^2 + 4)} = 6s\left[\frac{1}{(s^2 + 1)(s^2 + 4)}\right]}$$

Now express the function in the square brackets in partial fractions and then multiply by the factor $6s$ afterwards.

$\therefore \dfrac{6s}{(s^2 + 1)(s^2 + 4)} =$ ....

53

$$\boxed{2\left[\frac{s}{s^2+1}-\frac{s}{s^2+4}\right]}$$

For
$$\frac{6s}{(s^2+1)(s^2+4)} = 6s\left[\frac{1}{(s^2+1)(s^2+4)}\right]$$
$$= 6s\left[\frac{1/3}{s^2+1}+\frac{1/(-3)}{s^2+4}\right]$$
$$= 2s\left[\frac{1}{s^2+1}-\frac{1}{s^2+4}\right]$$
$$= 2\left[\frac{s}{s^2+1}-\frac{s}{s^2+4}\right]$$

All you now have to do is to put this back into the expression for $\bar{x}$ and then go on to find x.

What do you get?

54

$$\boxed{x = 5\cos t + \sin t - 2\cos 2t}$$

since
$$\bar{x} = \frac{3s}{s^2+1}+\frac{1}{s^2+1}+\frac{2s}{s^2+1}-\frac{2s}{s^2+4}$$
$$\therefore \bar{x} = \frac{5s}{s^2+1}+\frac{1}{s^2+1}-\frac{2s}{s^2+4}$$
$$\therefore \underline{x = 5\cos t + \sin t - 2\cos 2t}$$

Now you can do this one straight through:

Example 6. Solve: $\ddot{x} + 9x = \cos 2t$, given that at $t = 0, x = 1, \dot{x} = 3$.

Obtain the particular solution and then check with the detailed working given in frame 55.

55

$$x = \frac{4}{5}\cos 3t + \sin 3t + \frac{1}{5}\cos 2t$$

Here is the working:

$$\ddot{x} + 9x = \cos 2t \qquad x_0 = 1; \qquad x_1 = 3$$

$$(s^2.\bar{x} - sx_0 - x_1) + 9\bar{x} = \frac{s}{s^2+4}$$

$$\therefore s^2.\bar{x} - s - 3 + 9\bar{x} = \frac{s}{s^2+4}$$

$$\therefore (s^2+9)\bar{x} = s + 3 + \frac{s}{s^2+4}$$

$$\therefore \bar{x} = \frac{s}{s^2+9} + \frac{3}{s^2+9} + \frac{s}{(s^2+4)(s^2+9)}$$

Now

$$\frac{s}{(s^2+4)(s^2+9)} = s\left[\frac{1}{(s^2+4)(s^2+9)}\right]$$

$$= s\left[\frac{1/5}{s^2+4} + \frac{1/(-5)}{s^2+9}\right]$$

$$= \frac{s}{5}\left[\frac{1}{s^2+4} - \frac{1}{s^2+9}\right]$$

$$= \frac{1}{5}\left[\frac{s}{s^2+4} - \frac{s}{s^2+9}\right]$$

$$\therefore \bar{x} = \frac{s}{s^2+9} + \frac{3}{s^2+9} + \frac{1}{5}.\frac{s}{s^2+4} - \frac{1}{5}.\frac{s}{s^2+9}$$

$$= \frac{4}{5}.\frac{s}{s^2+9} + \frac{3}{s^2+9} + \frac{1}{5}.\frac{s}{s^2+4}$$

$$\therefore x = \frac{4}{5}\cos 3t + \sin 3t + \frac{1}{5}.\cos 2t$$

That's it. Now move on to frame 56.

56

Now let us do this one:

Example 7. Solve: $\ddot{x} + 3\dot{x} + 2x = 4t^2$, given that at $t = 0, x = 0, \dot{x} = 0$.
First go ahead and find an expression for $\bar{x}$

$\bar{x} =$ ....

57

$$\boxed{\bar{x} = \frac{8}{s^3(s+1)(s+2)}}$$

for:

$$\ddot{x} + 3\dot{x} + 2x = 4t^2$$

$$(s^2.\bar{x} - s.x_0 - x_1) + 3(s.\bar{x} - x_0) + 2\bar{x} = \frac{4.2!}{s^3}$$

$$x_0 = x_1 = 0$$

$$\therefore s^2.\bar{x} + 3s.\bar{x} + 2\bar{x} = \frac{8}{s^3}$$

$$\therefore (s^2 + 3s + 2)\bar{x} = \frac{8}{s^3}$$

$$\therefore \bar{x} = \frac{8}{s^3(s+1)(s+2)}$$

Now express this in partial fractions

$\bar{x} =$ ....

58

$$\bar{x} = \frac{4}{s^3} - \frac{6}{s^2} + \frac{7}{s} + \frac{1}{s+2} - \frac{8}{s+1}$$

Since, applying the tricks that are now familiar,

$$\bar{x} = \frac{1}{s^2}\left[\frac{8}{s(s+1)(s+2)}\right] \qquad = \frac{1}{s^2}\left[\frac{8/2}{s} + \frac{8/(-1)}{s+1} + \frac{8/2}{s+2}\right]$$

$$= \frac{1}{s^2}\left[\frac{4}{s} - \frac{8}{s+1} + \frac{4}{s+2}\right] \qquad = \frac{4}{s^3} - \frac{1}{s}\left[\frac{8}{s(s+1)} - \frac{4}{s(s+2)}\right]$$

$$= \frac{4}{s^3} - \frac{1}{s}\left[\frac{8}{s} - \frac{8}{s+1} - \frac{2}{s} + \frac{2}{s+2}\right] \qquad = \frac{4}{s^3} - \frac{6}{s^2} + \frac{8}{s(s+1)} - \frac{2}{s(s+2)}$$

$$= \frac{4}{s^3} - \frac{6}{s^2} + \frac{8}{s} - \frac{8}{s+1} - \frac{1}{s} + \frac{1}{s+2} \qquad = \frac{4}{s^3} - \frac{6}{s^2} + \frac{7}{s} + \frac{1}{s+2} - \frac{8}{s+1}$$

So now, $x = \mathrm{F}(t) = \ldots\ldots\ldots\ldots\ldots$

59

$$x = 2t^2 - 6t + 7 + e^{-2t} - 8e^{-t}$$

As you can see, the approach is much the same in each case, applying the standard transforms and inverse transforms – with a thorough practical knowledge of partial fractions – and working through the four distinct stages:

(i) Express the equation in terms of its Laplace transforms.
(ii) Insert the initial conditions.
(iii) Re-arrange to find the transform of the solution, i.e. $\bar{x}$.
(iv) Write the inverse transforms to obtain the solution, i.e. $x = \mathrm{F}(t)$.

Now, by way of revision, here is one more example for you to work through entirely on your own. When you have finished, turn on to the next frame and check the result.

Example 8. Solve the equation: $\ddot{x} - 4\dot{x} + 4x = \sin 3t$ given that at $t = 0, x = 0$ and $\dot{x} = 4$.

Take your time over it. Finish it completely before looking at the solution in frame 60.

60

Here is the working in full:

To solve: $\ddot{x} - 4\dot{x} + 4x = \sin 3t$, given that $t = 0, x = 0, \dot{x} = 4$.

$$(s^2.\bar{x} - s.x_0 - x_1) - 4(s.\bar{x} - x_0) + 4\bar{x} = \frac{3}{s^2+9}$$

$$x_0 = 0; \qquad x_1 = 4$$

$$\therefore s^2.\bar{x} - 4 - 4s.\bar{x} + 4\bar{x} = \frac{3}{s^2+9}$$

$$\therefore (s^2 - 4s + 4)\bar{x} = 4 + \frac{3}{s^2+9}$$

$$\therefore (s-2)^2\bar{x} = 4 + \frac{3}{s^2+9}$$

$$\therefore \bar{x} = \frac{4}{(s-2)^2} + \frac{3}{(s-2)^2(s^2+9)}$$

Now
$$\frac{3}{(s-2)^2(s^2+9)} = \frac{3}{s-2}\left[\frac{1}{(s-2)(s^2+9)}\right]$$

$$= \frac{3}{s-2}\left[\frac{1/13}{s-2} + \frac{-s/13 - 2/13}{s^2+9}\right]$$

$$= \frac{3}{13}\left[\frac{1}{(s-2)^2} - \frac{s+2}{(s-2)(s^2+9)}\right]$$

$$= \frac{3}{13}\left\{\frac{1}{(s-2)^2} - \left[\frac{4/13}{s-2} + \frac{-4s/13 + 5/13}{s^2+9}\right]\right\}$$

$$= \frac{3}{13}\cdot\frac{1}{(s-2)^2} - \frac{3}{169}\cdot\frac{4}{s-2} + \frac{12}{169}\cdot\frac{s}{s^2+9} - \frac{15}{169}\cdot\frac{1}{s^2+9}$$

$$\therefore \bar{x} = \frac{55}{13}\cdot\frac{1}{(s-2)^2} - \frac{3}{169}\cdot\frac{4}{s-2} + \frac{12}{169}\,\frac{s}{s^2+9} - \frac{15}{169}\cdot\frac{1}{s^2+9}$$

$$\therefore x = \frac{55}{13}t\,e^{2t} - \frac{3}{169}\left[4e^{2t} - 4\cos 3t + \frac{5}{3}\sin 3t\right]$$

There it is. If you made a slip, check through the solution carefully to see where you went wrong. It is all very straightforward, but needs care.

On then now to frame 61.

61

There are occasions, of course, when the quadratic coefficient of $\bar{x}$ cannot be expressed as straightforward linear factors. In such a case, we merely complete the square, converting the expression into $(s \pm k)^2 \pm a^2$ where the signs are determined by the particular expression.

Here is one, just to remind you:

Example 9. Solve: $\ddot{x} - 2\dot{x} + 5x = e^{2t}$ given that when $t = 0, x = 0, \dot{x} = 1$.

Proceeding as usual

$$(s^2\bar{x} - sx_0 - x_1) - 2(s\bar{x} - x_0) + 5\bar{x} = \frac{1}{s-2}$$

$x_0 = 0, x_1 = 1$

$$\therefore s^2\bar{x} - 1 - 2s\bar{x} + 5\bar{x} = \frac{1}{s-2}$$

$$\therefore (s^2 - 2s + 5)\bar{x} = 1 + \frac{1}{s-2}$$

$$\therefore \bar{x} = \frac{1}{s^2 - 2s + 5} + \frac{1}{(s-2)(s^2 - 2s + 5)}$$

In the first term, we now complete the square in the denominator, giving

$$\frac{1}{s^2 - 2s + 5} = \ldots\ldots\ldots\ldots$$

62

$$\boxed{\frac{1}{s^2 - 2s + 5} = \frac{1}{(s-1)^2 + 4}}$$

We are happy to accept this in this form since $\dfrac{1}{(s-1)^2 + 4}$ is merely $\dfrac{1}{s^2 + 4}$ with s replaced by $(s - 1)$, indicating an extra factor e^t in the final function. (Remember?)

The second term $\dfrac{1}{(s-2)(s^2 - 2s + 5)}$ is now expressed in partial fractions,

giving $\ldots\ldots\ldots\ldots$

63

$$\boxed{\frac{1}{(s-2)(s^2-2s+5)} = \frac{1}{5}\cdot\frac{1}{s-2} - \frac{1}{5}\cdot\frac{s}{s^2-2s+5}}$$

$$\therefore \bar{x} = \frac{1}{(s-1)^2+4} + \frac{1}{5}\cdot\frac{1}{s-2} - \frac{1}{5}\cdot\frac{s}{s^2-2s+5}$$

$$= \frac{1}{(s-1)^2+4} + \frac{1}{5}\cdot\frac{1}{s-2} - \frac{1}{5}\left[\frac{s-1}{(s-1)^2+4} + \frac{1}{(s-1)^2+4}\right]$$

$$\therefore \bar{x} = \frac{1}{(s-1)^2+4} + \frac{1}{5}\cdot\frac{1}{s-2} - \frac{1}{5}\cdot\frac{s-1}{(s-1)^2+4} - \frac{1}{5}\cdot\frac{1}{(s-1)^2+4}$$

Collecting the first and last terms –

$$\bar{x} = \frac{4}{5}\cdot\frac{1}{(s-1)^2+4} + \frac{1}{5}\cdot\frac{1}{s-2} - \frac{1}{5}\cdot\frac{s-1}{(s-1)^2+4}$$

$\therefore x =$ ....

64

$$\boxed{x = \frac{1}{5}\left[e^{2t} - e^{t}(\cos 2t - 2\sin 2t)\right]}$$

since (i) $\mathcal{L}^{-1}\left\{\dfrac{1}{s^2+4}\right\} = \dfrac{1}{2}\mathcal{L}^{-1}\left\{\dfrac{2}{s^2+4}\right\} = \dfrac{1}{2}\sin 2t$

$$\therefore \mathcal{L}^{-1}\left\{\frac{1}{(s-1)^2+4}\right\} = \frac{1}{2}\sin 2t\,.\,e^{t}$$

(ii) $\mathcal{L}^{-1}\left\{\dfrac{1}{s-2}\right\} = e^{2t}$

(iii) $\mathcal{L}^{-1}\left\{\dfrac{s}{s^2+4}\right\} = \cos 2t \qquad \therefore \mathcal{L}^{-1}\left\{\dfrac{s-1}{(s-1)^2+4}\right\} = \cos 2t\,.\,e^{t}$

$$\therefore x = \frac{4}{5}\cdot\frac{1}{2}e^{t}\sin 2t + \frac{1}{5}e^{2t} - \frac{1}{5}e^{t}\cos 2t$$

$$\therefore x = \frac{1}{5}\left[e^{2t} - e^{t}(\cos 2t - 2\sin 2t)\right]$$

Once again, it all depends on Partial Fractions.

On to frame 65.

65

Solution of third order differential equations

We have dealt in some detail with the solution of first and second order equations using the Laplace transform method. Third order equations are solved in just the same way, though the working may sometimes be a little more lengthy and tedious. Let us work through one example.

Example 1. Solve the equation: $\dddot{x} + \ddot{x} + \dot{x} + x = 4$, given that at $t = 0, x = \dot{x} = 0$, $\ddot{x} = 3$.

Follow the usual steps:

(1) Express the equation in terms of Laplace transforms, giving

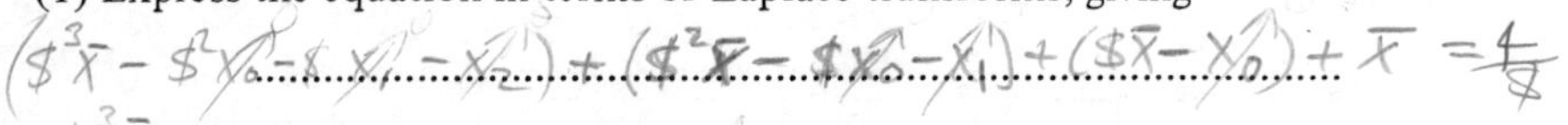

66

$$(s^3\bar{x} - s^2x_0 - sx_1 - x_2) + (s^2\bar{x} - sx_0 - x_1) + (s\bar{x} - x_0) + \bar{x} = \frac{4}{s}$$

(ii) Insert the initial conditions

$$x_0 = 0, \qquad x_1 = 0, \qquad x_2 = 3$$

The equation now becomes – ....

67

$$(s^3\bar{x} - 3) + s^2\bar{x} + s\bar{x} + \bar{x} = \frac{4}{s}$$

$$\therefore (s^3 + s^2 + s + 1)\bar{x} - 3 = \frac{4}{s}$$

$$(s^3 + s^2 + s + 1)\bar{x} = 3 + \frac{4}{s} = \frac{3s + 4}{s}$$

(iii) Now factorize the coefficient of x, i.e. $(s^3 + s^2 + s + 1)$.

$(s^3 + s^2 + s + 1) =$ ....

68

$$\boxed{(s+1)(s^2+1)}$$

for
$$s^3+s^2+s+1 = s^2(s+1)+(s+1)$$
$$= (s+1)(s^2+1)$$
$$\therefore (s+1)(s^2+1)\,\overline{x} = \frac{3s+4}{s}$$
$$\therefore \overline{x} = \frac{3s+4}{s(s+1)(s^2+1)}$$
$$x = \frac{3(s+1)}{s(s+1)(s^2+1)} + \frac{1}{s(s+1)(s^2+1)}$$
$$= \frac{3}{s(s^2+1)} + \frac{1}{s(s+1)(s^2+1)}$$

Now we resort to partial fractions once again.

$$\frac{3}{s(s^2+1)} = 3\cdot\frac{1}{s(s^2+1)} = \text{..}$$

69

$$\boxed{\frac{3}{s(s^2+1)} = 3\left[\frac{1}{s} - \frac{s}{s^2+1}\right]}$$

Now express the second term in partial fractions

$$\frac{1}{s(s+1)(s^2+1)} = \frac{1}{s+1}\left[\frac{1}{s(s^2+1)}\right]$$
$$= \frac{1}{s+1}\left[\frac{1}{s} - \frac{s}{s^2+1}\right] \quad \text{using the result above.}$$
$$= \frac{1}{s(s+1)} - \frac{s}{(s+1)(s^2+1)}$$

and expressing each of these in partial fractions, we get

$$\frac{1}{s(s+1)(s^2+1)} = \text{....................................}$$

70

$$\boxed{\frac{1}{s}-\frac{1}{2}\cdot\frac{1}{s+1}-\frac{1}{2}\cdot\frac{s}{s^2+1}-\frac{1}{2}\cdot\frac{1}{s^2+1}}$$

for
$$\frac{1}{s(s+1)(s^2+1)}=\frac{1}{s(s+1)}-\frac{s}{(s+1)(s^2+1)}$$
$$=\left[\frac{1}{s}+\frac{(-1)}{s+1}\right]-\left[\frac{-1/2}{s+1}+\frac{s/2+1/2}{s^2+1}\right]$$
$$=\frac{1}{s}-\frac{1}{s+1}+\frac{1}{2}\cdot\frac{1}{s+1}-\frac{1}{2}\cdot\frac{s+1}{s^2+1}$$
$$=\frac{1}{s}-\frac{1}{2}\cdot\frac{1}{s+1}-\frac{1}{2}\cdot\frac{s}{s^2+1}-\frac{1}{2}\cdot\frac{1}{s^2+1}$$

So, collecting together the two sets of partial fractions, we have

$\bar{x}=$ ....

71

$$\boxed{\bar{x}=3\left[\frac{1}{s}-\frac{s}{s^2+1}\right]+\left[\frac{1}{s}-\frac{1}{2}\cdot\frac{1}{s+1}-\frac{1}{2}\cdot\frac{s}{s^2+1}-\frac{1}{2}\cdot\frac{1}{s^2+1}\right]}$$

which can now be tidied up to read
$$\bar{x}=\frac{4}{s}-\frac{1}{2}\cdot\frac{1}{s+1}-\frac{7}{2}\cdot\frac{s}{s^2+1}-\frac{1}{2}\cdot\frac{1}{s^2+1}$$

Finally, take inverse transforms, so that

$x=$ ....

72

$$\boxed{x=4-\frac{e^{-t}}{2}-\frac{7}{2}\cos t-\frac{1}{2}\sin t}$$

and that is that.

As you see, apart from a more extensive use of partial fractions, the solution of a third order equation is really no different from those we have done before.

Here is another:

Example 2. Solve: $\dddot{x}-2\ddot{x}-\dot{x}+2x=2+t$ given that at $t=0, x=0, \dot{x}=1, \ddot{x}=0$.

Work right through the solution and then check with the result set out in detail in the next frame.

73

Here is the solution:

Solve: $\dddot{x} - 2\ddot{x} - \dot{x} + 2x = 2 + t$ with $x = 0, \dot{x} = 1, \ddot{x} = 0$, at $t = 0$.

$$(s^3\bar{x} - s^2x_0 - sx_1 - x_2) - 2(s^2\bar{x} - sx_0 - x_1) - (s\bar{x} - x_0) + 2\bar{x} = \frac{2}{s} + \frac{1}{s^2}$$

Initial conditions: $x_0 = 0, x_1 = 1, x_2 = 0$.

$$\therefore s^3\bar{x} - s - 2s^2\bar{x} + 2 - s\bar{x} + 2\bar{x} = \frac{2s+1}{s^2}$$

$$(s^3 - 2s^2 - s + 2)\bar{x} - s + 2 = \frac{2s+1}{s^2}$$

$$(s-2)(s^2-1)\bar{x} = s - 2 + \frac{2s+1}{s^2}$$

$$\therefore \bar{x} = \frac{s-2}{(s-2)(s^2-1)} + \frac{2s+1}{s^2(s-2)(s^2-1)}$$

$$= \frac{1}{s^2-1} + \frac{1}{s}\left[\frac{2s+1}{s(s-1)(s+1)(s-2)}\right]$$

$$\therefore \bar{x} = \frac{1}{s^2-1} + \frac{1}{s}\left[\frac{1}{2}\cdot\frac{1}{s} - \frac{3}{2}\cdot\frac{1}{s-1} + \frac{1}{6}\cdot\frac{1}{s+1} + \frac{5}{6}\cdot\frac{1}{s-2}\right]$$

$$\therefore \bar{x} = \frac{1}{s^2-1} + \frac{1}{2}\cdot\frac{1}{s^2} - \frac{3}{2}\cdot\frac{1}{s(s-1)} + \frac{1}{6}\cdot\frac{1}{s(s+1)} + \frac{5}{6}\cdot\frac{1}{s(s-2)}]$$

$$= \frac{1}{s^2-1} + \frac{1}{2}\cdot\frac{1}{s^2} - \frac{3}{2}\left[-\frac{1}{s} + \frac{1}{s-1}\right] + \frac{1}{6}\left[\frac{1}{s} - \frac{1}{s+1}\right] + \frac{5}{6}\left[-\frac{1}{2}\cdot\frac{1}{s} + \frac{1}{2}\cdot\frac{1}{s-2}\right]$$

$$= \frac{1}{s^2-1} + \frac{1}{2}\cdot\frac{1}{s^2} + \frac{3}{2}\cdot\frac{1}{s} + \frac{3}{2}\cdot\frac{1}{s-1} + \frac{1}{6}\cdot\frac{1}{s} - \frac{1}{6}\cdot\frac{1}{s+1} - \frac{5}{12}\cdot\frac{1}{s} + \frac{5}{12}\cdot\frac{1}{s-2}$$

$$= \frac{1}{s^2-1} + \frac{1}{2}\cdot\frac{1}{s^2} + \frac{5}{4}\cdot\frac{1}{s} - \frac{3}{2}\cdot\frac{1}{s-1} - \frac{1}{6}\cdot\frac{1}{s+1} + \frac{5}{12}\cdot\frac{1}{s-2}$$

$$\therefore x = \sinh t + \frac{t}{2} + \frac{5}{4} - \frac{3e^t}{2} - \frac{e^{-t}}{6} + \frac{5}{12}e^{2t}$$

or, since $\sinh t = \frac{1}{2}(e^t - e^{-t})$, the result could be written

$$x = \frac{5}{4} + \frac{t}{2} - e^t - \frac{2e^{-t}}{3} + \frac{5}{12}e^{2t}$$

74

All that now remains is the Test Exercise. Before working through it, check down the revision summary listed below. If there is any point about which you are not completely clear, go back and work through the relative part of the programme again. There is no rush: it is important to be sure of every point.

Revision Summary

1. *Transforms of differential coefficients of* $x = F(t)$

$$\mathcal{L}\{F^n(t)\} = s^n.f(s) - s^{n-1}.\,F(0) - s^{n-2}.F^1(0) - \ldots - F^{n-1}(0).$$

where

$$f(s) = \mathcal{L}\{F(t)\}$$

$$F^n(t) = \frac{d^n x}{dt^n}$$

Alternative notation:

$$\mathcal{L}\left\{\frac{d^n x}{dt^n}\right\} = s^n.\bar{x} - s^{n-1}x_0 - s^{n-2}x_1 - \ldots - x_{n-1}.$$

$$\mathcal{L}\{x\} = \bar{x}$$

$$\mathcal{L}\left\{\frac{dx}{dt}\right\} = s.\bar{x} - x_0$$

$$\mathcal{L}\left\{\frac{d^2x}{dt^2}\right\} = s^2.\bar{x} - s.x_0 - x_1$$

$$\mathcal{L}\left\{\frac{d^3x}{dt^3}\right\} = s^3.\bar{x} - s^2.x_0 - s.x_1 - x_2 \qquad \text{etc.}$$

2. *Solution of differential equations*

(i) Rewrite the equation in terms of its Laplace transforms.
(ii) Insert initial conditions.
(iii) Re-arrange the equation in s to give the transform of the solution ($\bar{x}$).
(iv) Determine the inverse transforms to obtain the solution $x = F(t)$.

When you are ready, move on the Text Exercise in frame 75.

75

Work through all the problems below. Take care and all the time you need. Accuracy is always important and especially with the partial fractions.

Test Exercise–III

Solve the following differential equations by the method of Laplace transforms.

1. $2\dot{x} + 5x = e^{-2t}$ given that at $t = 0, x = 1$.
2. $4\dot{x} - 3x = \sin 2t$ given that at $t = 0, x = 1/4$.
3. $\ddot{x} - 5\dot{x} + 6x = 0$ given that at $t = 0, x = 0, \dot{x} = 5$.
4. $\ddot{x} + 6\dot{x} + 5x = 2$ given that at $t = 0, x = 1, \dot{x} = 2$.
5. $\ddot{x} - 4x = \cos 2t$ given that at $t = 0, x = 2, \dot{x} = 3$.
6. $\ddot{x} - 2\dot{x} + x = t\,e^{t}$ given that at $t = 0, x = 1, \dot{x} = 0$.
7. $\ddot{x} - 4\dot{x} + 5x = e^{3t}$ given that at $t = 0, x = 0, \dot{x} = 2$.
8. $\dddot{x} - \ddot{x} + \dot{x} - x = e^{2t}$ given that at $t = 0, x = 0, \dot{x} = 1, \ddot{x} = 0$.

Now on to the next programme.

Further Problems III

Solve the following differential equations:

1. $\ddot{x} - 3\dot{x} + 2x = \sin t$, given that at $t = 0, x = 0, \dot{x} = 1$.
2. $\ddot{x} + 4x = t$, given that at $t = 0, x = 1, \dot{x} = 0$.
3. $\ddot{x} + 4\dot{x} + 3x = 12$ with $x = 7, \dot{x} = 1$ at $t = 0$.
4. $\ddot{y} + 4\dot{y} + 8y = \cos 2t$, with $y = 2, \dot{y} = 1$ at $t = 0$.
5. $\ddot{x} - 2\dot{x} + x = e^{t}$, given that at $t = 0, x = -2, \dot{x} = -3$.
6. $\ddot{y} + 2\dot{y} + y = \sin t$, given that $y = 3, \dot{y} = 1$ at $t = 0$.
7. $\ddot{x} + 3\dot{x} + 2x = 4t^2$ with $x = \dot{x} = 0$ at $t = 0$.
8. $\ddot{x} + 4\dot{x} + 5x = 8 \sin t$, given that at $t = 0, x = 4, \dot{x} = 0$.
9. $\ddot{y} + 4\dot{y} + 4y = t^2 + e^{-2t}$, given that at $t = 0, y = \frac{1}{2}, \dot{y} = 0$.
10. $\dddot{x} + \ddot{x} + \dot{x} + x = \cos t$ with $x = \dot{x} = \ddot{x} = 0$ at $t = 0$.

Find the general solutions of equations Nos. 11, 12, 13:

11. $\ddot{y} - 9y = \cosh 3t + t^2$.
12. $\ddot{y} - \dot{y} - 6y = t\,e^{3t}$.
13. $\ddot{y} + 3\dot{y} + 2y = e^{-t} \cos t$.
14. Solve the equation

$$\ddot{y} - 3\dot{y} + 2y = e^{2t}(3t^2 + 2 \cos t)$$

given that $y = 1, \dot{y} = 8$ when $t = 0$.

15. The equation of motion of a mass moving in a straight line is given by

$$\ddot{x} + 2k\dot{x} + (k^2 + n^2)x = 0 \qquad (n \neq 0)$$

x being the displacement from an initial point at time t. Find x in terms of t, given that $x = a$, $\dot{x} = 0$ at $t = 0$.

16. An e.m.f. $E \sin \omega t$ acts at $t = 0$ on an inductive resistance (L,R) and the initial condition is zero. Show that at time t,

$$i = \frac{E}{\sqrt{(R^2 + L^2\omega^2)}}\left\{e^{-Rt/L} \sin\theta + \sin(\omega t - \theta)\right\} \text{ where } \tan\theta = L\omega/R.$$

17. The function $v = F(t)$ satisfies the equation

$$\frac{d^2v}{dt^2} - \frac{dv}{dt} - 2v = -12\,e^{-2t}(\cos t - 4 \sin t)$$

and the initial conditions that at $t = 0$, $v = 0$ and $\dfrac{dv}{dt} = 3$. Show that the Laplace transform of $F(t)$ is

$$f(s) = \frac{3(s^2 + 13)}{(s + 1)(s - 2)(s^2 + 4s + 5)} \text{ and hence determine } F(t).$$

18. The equation of motion of a mass suspended from a spring when the support is subjected to vertical vibrations $F(t) = \sin 4t$ is given by

$$\ddot{y} + 8\dot{y} + 32y = 32 \sin 4t.$$

At $t = 0, y = \dot{y} = 0$. Find an expression for y in terms of t.

19. Show that if θ satisfies the equation

$$\ddot{\theta} + 2k\dot{\theta} + n^2\theta = 0 \qquad (k < n)$$

and if, when $t = 0, \theta = a, \dot{\theta} = 0$,then

$$\theta = e^{-kt}\left(a \cos pt + \frac{ka}{p} \sin pt\right)$$

where $p^2 = n^2 - k^2$.

20. A unit mass P moves in a straight line through a point 0 so that at time t:

(i) its distance from 0 is x,
(ii) its velocity is v,
(iii) it is acted on by a force w^2x towards 0,
(iv) it experiences a resistance $2kv$,
(v) it is subjected to a force e^{-kt} cos pt.

Show that $\ddot{x} + 2k\dot{x} + w^2x = e^{-kt} \cos pt$ and determine x in terms of t, when $p = n$ where $n^2 = w^2 - k^2$.

Programme 4

SIMULTANEOUS EQUATIONS

1

Introduction

In the previous programme, we applied our knowledge of Laplace transforms to the solution of numerous differential equations with constant coefficients. The method is particularly useful when the conditions at $t = 0$ are known, for these are substituted early in the proceedings and the particular solution of the equation relating to the given initial conditions is automatically obtained.

Many situations in science and engineering give rise to simultaneous differential equations and we are now going to apply our techniques to the solution of such sets of equations.

The general approach is almost the same as before:

(i) Express the equations in terms of Laplace transforms,
(ii) Simplify the equations in $\bar{x}$ and $\bar{y}$,
(iii) Solve for $\bar{x}$ and $\bar{y}$ by normal algebraic methods,
(iv) Determine the inverse transforms of $\bar{x}$ and $\bar{y}$.

The examples have been chosen to show some useful steps in solution, so let us make a start.

On then to Frame 2.

2

Example 1. To solve the equations

$$\left.\begin{aligned} \frac{dx}{dt} + \frac{dy}{dt} + 2x &= 0 \\ \frac{dx}{dt} + 4\frac{dy}{dt} - 2x &= e^{-2t} \end{aligned}\right\} \quad \text{given that at } t = 0,\ x = 2,\ y = 0.$$

(i) Using the usual notation, i.e. letting $\bar{x}$ denote $\mathcal{L}\{x\}$
and $\bar{y}$ denote $\mathcal{L}\{y\}$

we first express the two equations in terms of the Laplace transforms, giving

....

....

3

$$(s\bar{x} - x_0) + (s\bar{y} - y_0) + 2\bar{x} = 0$$
$$(s\bar{x} - x_0) + 4(s\bar{y} - y_0) - 2\bar{x} = \frac{1}{s+2}$$

(ii) Now insert the initial conditions in both equations. In this case, $x_0 = 2$ and $y_0 = 0$

so the equations become ....

and ....

4

$$(s\bar{x} - 2) + s\bar{y} + 2\bar{x} = 0$$
$$(s\bar{x} - 2) + 4s\bar{y} - 2\bar{x} = \frac{1}{s+2}$$

Collecting up terms in $\bar{x}$ and $\bar{y}$ gives

....

....

5

$$(s + 2)\bar{x} + s\bar{y} = 2$$
$$(s - 2)\bar{x} + 4s\bar{y} = 2 + \frac{1}{s+2}$$

So we have a pair of straightforward simultaneous equations in $\bar{x}$ and $\bar{y}$.

(iii) We can solve these algebraically by eliminating one variable while we concentrate on the other. By multiplying the first equation by 4, we can eliminate $\bar{y}$ and so obtain an expression for $\bar{x}$.

Do that, and then check with the next frame.

$\bar{x} =$ ....

6

$$\boxed{\bar{x} = \frac{6}{3s+10} - \frac{1}{(s+2)(3s+10)}}$$

for:
$$4(s+2)\bar{x} + 4s\bar{y} = 8$$
$$(s-2)\bar{x} + 4s\bar{y} = 2 + \frac{1}{s+2}$$

Subtract $\quad \therefore\ (3s+10)\bar{x} = 6 - \dfrac{1}{s+2}$

$$\therefore\ \bar{x} = \frac{6}{3s+10} - \frac{1}{(s+2)(3s+10)}$$

Now we come to step (iv). The first term $\dfrac{6}{3s+10}$ can be written as $\dfrac{2}{s+10/3}$ which is a form we recognize, but the second term must be expressed in partial fractions.

Applying the usual 'cover-up' rule

$$\frac{1}{(s+2)(3s+10)} = \ldots\ldots\ldots\ldots$$

7

$$\boxed{\frac{1}{4}\cdot\frac{1}{s+2} - \frac{1}{4}\cdot\frac{1}{s+10/3}}$$

since
$$\frac{1}{(s+2)(3s+10)} = \frac{1/4}{s+2} + \frac{1/(-4/3)}{3s+10}$$
$$= \frac{1}{4}\cdot\frac{1}{s+2} - \frac{3}{4}\cdot\frac{1}{3s+10}$$
$$= \frac{1}{4}\cdot\frac{1}{s+2} - \frac{1}{4}\cdot\frac{1}{s+10/3}$$

So the complete expression for $\bar{x}$ becomes

$$\bar{x} = \frac{2}{s+10/3} - \frac{1}{4}\cdot\frac{1}{s+2} + \frac{1}{4}\cdot\frac{1}{s+10/3} = \frac{9}{4}\cdot\frac{1}{s+10/3} - \frac{1}{4}\cdot\frac{1}{s+2}$$

Finally, taking inverse transforms, this gives

$$x = \ldots\ldots\ldots\ldots$$

8

$$x = \frac{9}{4}e^{-10t/3} - \frac{1}{4}e^{-2t}$$

And so we have the solution for x. But that is only half the story: we now have to determine y.

To find y, we usually go back and make a fresh start by eliminating $\bar{x}$.

$$(s + 2)\bar{x} + s\bar{y} = 2 \qquad \ldots \text{ (i)}$$

$$(s - 2)\bar{x} + 4s\bar{y} = 2 + \frac{1}{s + 2} \qquad \ldots \text{ (ii)}$$

Multiply (i) by $(s - 2)$ and (ii) by $(s + 2)$ and proceed to find an expression for $\bar{y}$.

$\bar{y}$ = ...

9

$$\bar{y} = \frac{3}{s(s + 10/3)}$$

for we have:

$$(s + 2)(s - 2)\,\bar{x} + s(s - 2)\,\bar{y} = 2(s - 2)$$

$$(s + 2)(s - 2)\,\bar{x} + 4s(s + 2)\,\bar{y} = 2(s + 2) + 1$$

Subtract $\quad \therefore (3s^2 + 10s)\,\bar{y} = 9$

$$\therefore \bar{y} = \frac{9}{s(3s + 10)} = \frac{3}{s(s + 10/3)}$$

Expressing this in partial fractions, we obtain

$\bar{y}$ = ...

10

$$\boxed{\bar{y} = \frac{9}{10}\cdot\frac{1}{s} - \frac{9}{10}\cdot\frac{1}{s + 10/3}}$$

The inverse transforms of this, of course, immediately give the solution.

$$\therefore y = \frac{9}{10}(1 - e^{-10t/3})$$

So, collecting the two solutions together, we have

$$x = \frac{1}{4}\left\{9e^{-10t/3} - e^{-2t}\right\}$$

$$y = \frac{9}{10}\left\{1 - e^{-10t/3}\right\}.$$

Let us work through another example in much the same way.

11

Example 2. Solve the equations

$$\left.\begin{aligned} 2\dot{x} - x + \dot{y} + y &= 5 \sin t \\ 3\dot{x} - x + 2\dot{y} + y &= e^{t} \end{aligned}\right\} \quad \text{given that at } t = 0, x = y = 0.$$

Express both equations in Laplace transforms, insert the initial conditions and tidy up the resulting expressions, giving:

....

....

12

$$(2s-1)\bar{x}+(s+1)\bar{y}=\frac{5}{s^2+1}$$

$$(3s-1)\bar{x}+(2s+1)\bar{y}=\frac{1}{s-1}$$

for:
$$2(s\bar{x}-x_0)-\bar{x}+(s\bar{y}-y_0)+\bar{y}=\frac{5}{s^2+1}$$

$$3(s\bar{x}-x_0)-\bar{x}+2(s\bar{y}-y_0)+\bar{y}=\frac{1}{s-1}$$

Initial conditions: $x_0=y_0=0$

$$\therefore \begin{cases} 2s\bar{x}-\bar{x}+s\bar{y}+\bar{y}=\dfrac{5}{s^2+1} \\ 3s\bar{x}-\bar{x}+2s\bar{y}+\bar{y}=\dfrac{1}{s-1} \end{cases}$$

$$\therefore \begin{cases} (2s-1)\bar{x}+(s+1)\bar{y}=\dfrac{5}{s^2+1} & \quad \ldots \text{(i)} \\ (3s-1)\bar{x}+(2s+1)\bar{y}=\dfrac{1}{s-1} & \quad \ldots \text{(ii)} \end{cases}$$

Now, to find $\bar{x}$, we eliminate $\bar{y}$ and so obtain

$\bar{x}=$ ....

13

$$\boxed{\bar{x} = \frac{5(2s+1)}{s(s-2)(s^2+1)} - \frac{s+1}{s(s-1)(s-2)}}$$

since:
$$(2s-1)(2s+1)\bar{x} + (s+1)(2s+1)\bar{y} = \frac{5(2s+1)}{s^2+1}$$

$$(s+1)(3s-1)\bar{x} + (s+1)(2s+1)\bar{y} = \frac{s+1}{s-1}$$

Subtract
$$\therefore (s^2 - 2s)\bar{x} = \frac{5(2s+1)}{s^2+1} - \frac{s+1}{s-1}$$

$$\therefore \bar{x} = \frac{5(2s+1)}{s(s-2)(s^2+1)} - \frac{s+1}{s(s-1)(s-2)}$$

Let us deal with these terms one at a time.
In partial fractions, the first term

$$\frac{5(2s+1)}{s(s-2)(s^2+1)} \equiv \ldots\ldots\ldots\ldots$$

14

$$\boxed{\frac{5}{2}\left\{\frac{1}{s-2} - \frac{1}{s}\right\} - \frac{5}{s^2+1}}$$

for:
$$\frac{5(2s+1)}{s(s-2)(s^2+1)} \equiv 5\left\{\frac{1/(-2)}{s} + \frac{5/10}{s-2} + \frac{?}{s^2+1}\right\}$$

$$= 5\left\{\frac{1}{2}\left(\frac{1}{s-2} - \frac{1}{s}\right) + \frac{?}{s^2+1}\right\}$$

$$= 5\left\{\frac{1}{s^2-2s} - \frac{1}{s^2+1}\right\}$$

$\therefore$ First term
$$= \frac{5}{2}\left\{\frac{1}{s-2} - \frac{1}{s}\right\} - \frac{5}{s^2+1}. \qquad \ldots \text{(i)}$$

And the second term

$$\frac{s+1}{s(s-1)(s-2)} \equiv \ldots\ldots\ldots\ldots$$

15

$$\boxed{\frac{1}{2}\cdot\frac{1}{s}-\frac{2}{s-1}+\frac{3}{2}\cdot\frac{1}{s-2}}$$

Applying the 'cover-up' rule gives that without any trouble. Collecting our two expressions together, we now have

$$\bar{x}=\frac{5}{2}\left(\frac{1}{s-2}-\frac{1}{s}\right)-\frac{5}{s^2+1}-\left(\frac{1}{2}\cdot\frac{1}{s}-\frac{2}{s-1}+\frac{3}{2}\cdot\frac{1}{s-2}\right)$$

$$=\frac{1}{s-2}+\frac{2}{s-1}-\frac{3}{s}-\frac{5}{s^2+1}$$

$$\therefore\ x= \ldots\ldots\ldots\ldots\ldots\ldots$$

16

$$\boxed{x=\mathrm{e}^{2t}+2\mathrm{e}^{t}-3-5\sin t}$$

Now, to find y, go back to equations (i) and (ii) in frame 12 and eliminate $\bar{x}$ in the same way, to obtain an expression for $\bar{y}$.

$$\bar{y}= \ldots\ldots\ldots\ldots\ldots\ldots$$

Then check with the next frame.

17

$$\boxed{\bar{y}=\frac{2s-1}{s(s-1)(s-2)}-\frac{5(3s-1)}{s(s-1)(s^2+1)}}$$

Here is the working for you to check

$$\begin{cases}(2s-1)\bar{x}+(s+1)\bar{y}=\dfrac{5}{s^2+1}\\[2ex](3s-1)\bar{x}+(2s+1)\bar{y}=\dfrac{1}{s-1}\end{cases}$$

$$\therefore\ \begin{cases}(2s-1)(3s-1)\bar{x}+(s+1)(3s-1)\bar{y}=\dfrac{5(3s-1)}{s^2+1}\\[2ex](2s-1)(3s-1)\bar{x}+(2s-1)(2s+1)\bar{y}=\dfrac{2s-1}{s-1}\end{cases}$$

Subtract $\qquad \therefore\ (s^2-2s)\bar{y}=\dfrac{2s-1}{s-1}-\dfrac{5(3s-1)}{s^2+1}$

$$\therefore\ \bar{y}=\frac{2s-1}{s(s-1)(s-2)}-\frac{5(3s-1)}{s(s-2)(s^2+1)}$$

Now express each of these terms in partial fractions. What do you get?

18

$$\frac{2s-1}{s(s-1)(s-2)} = -\frac{1}{2}\cdot\frac{1}{s} - \frac{1}{s-1} + \frac{3}{2}\cdot\frac{1}{s-2}$$

$$\frac{5(3s-1)}{s(s-2)(s^2+1)} = 5\left\{\frac{1}{2}\cdot\frac{1}{s} + \frac{1}{2}\cdot\frac{1}{s-2} - \frac{s}{s^2+1} - \frac{1}{s^2+1}\right\}$$

The first is easily obtained by the 'cover-up' rule.

The second is obtained like this:–

$$\frac{5(3s-1)}{s(s-2)(s^2+1)} = 5\left\{\frac{-1/(-2)}{s} + \frac{5/10}{s-2} + \frac{?}{s^2+1}\right\}$$

$$= 5\left\{\frac{1}{2}\left(\frac{1}{s} + \frac{1}{s-2}\right) + \frac{?}{s^2+1}\right\}$$

$$= 5\left\{\frac{s-1}{s^2-2s} + \frac{?}{s^2+1}\right\}$$

The missing numerator is of the form (As + B) and by multiplying up can be seen to be $(-s-1)$

$$\therefore \frac{5(3s-1)}{s(s-2)(s^2+1)} = 5\left\{\frac{1}{2}\cdot\frac{1}{s} + \frac{1}{2}\cdot\frac{1}{s-2} - \frac{s+1}{s^2+1}\right\}$$

$$= 5\left\{\frac{1}{2}\cdot\frac{1}{s} + \frac{1}{2}\cdot\frac{1}{s-2} - \frac{s}{s^2+1} - \frac{1}{s^2+1}\right\}$$

If you agree with these, move on to frame 19.

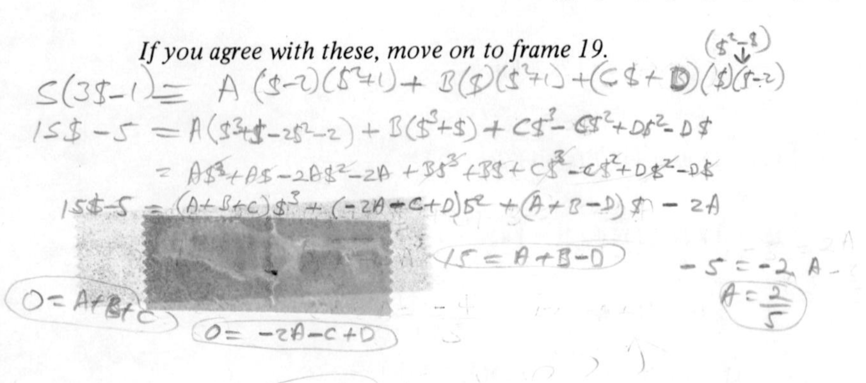

19

Collecting the two sets of partial fractions together, we have

$$\bar{y} = \left(-\frac{1}{2}\cdot\frac{1}{s} - \frac{1}{s-1} + \frac{3}{2}\cdot\frac{1}{s-2}\right) - \left(5\left\{\frac{1}{2}\cdot\frac{1}{s} + \frac{1}{2}\cdot\frac{1}{s-2} - \frac{s}{s^2+1} - \frac{1}{s^2+1}\right\}\right)$$

$$= -\frac{1}{2}\cdot\frac{1}{s} - \frac{1}{s-1} + \frac{3}{2}\cdot\frac{1}{s-2} - \frac{5}{2}\cdot\frac{1}{s} - \frac{5}{2}\cdot\frac{1}{s-2} + \frac{5s}{s^2+1} + \frac{5}{s^2+1}:$$

which can now be tidied up to give

$$\bar{y} = -\frac{3}{s} - \frac{1}{s-1} - \frac{1}{s-2} + \frac{5s}{s^2+1} + \frac{5}{s^2+1}$$

so that

$y =$ .... $-3 - e^{t} - e^{2t} + 5\cos t + 5\sin t$

20

$$\boxed{y = -3 - e^{t} - e^{2t} + 5\cos t + 5\sin t}$$

The solution to the whole problem is thus:

$$\begin{cases} \underline{x = e^{2t} + 2e^{t} - 3 - 5\sin t} \\ \underline{y = 5\cos t + 5\sin t - 3 - e^{t} - e^{2t}} \end{cases}$$

Simultaneous equations are all solved in much the same way. Here is one for you to do on your own.

Example 3. Solve the equations

$$\left.\begin{aligned} 4\dot{y} - 2\dot{x} + 10y - 5x &= 0 \\ \dot{x} - 18y + 15x &= 10 \end{aligned}\right\} \quad \text{given that at } t = 0, x = 4, y = 2.$$

You will know the steps to take, but always keep your eyes open for any 'short cuts' that will save you lengthy working. — — — That is a hint, so keep it in mind.

Off you go then. Complete the solutions and then check with the working set out in frame 21.

$$x = \frac{7}{3}\,e^{-6t} + \frac{5}{3}; \qquad y = \frac{x}{2} = \frac{7}{6}\,e^{-6t} + \frac{5}{6}$$

Here is the solution in detail:

$$\left.\begin{aligned} &4\dot{y} - 2\dot{x} + 10y - 5x = 0 \\ &\dot{x} - 18y + 15x = 10 \end{aligned}\right\} \qquad t = 0, x = 4, y = 2.$$

$$\left.\begin{aligned} &4(s\bar{y} - y_0) - 2(s\bar{x} - x_0) + 10\bar{y} - 5\bar{x} = 0 \\ &(s\bar{x} - x_0) - 18\bar{y} + 15\bar{x} = \frac{10}{s} \end{aligned}\right\} \qquad x_0 = 4; y_0 = 2$$

$$\therefore \left\{\begin{aligned} &4s\bar{y} - 8 - 2s\bar{x} + 8 + 10\bar{y} - 5\bar{x} = 0 \\ &s\bar{x} - 4 - 18\bar{y} + 15\bar{x} = \frac{10}{s} \end{aligned}\right.$$

$$\therefore \left\{\begin{aligned} &(4s + 10)\bar{y} - (2s + 5)\bar{x} = 0 &\quad \ldots \quad \text{(i)} \\ &(s + 15)\bar{x} - 18\bar{y} = 4 + \frac{10}{s} &\quad \ldots \quad \text{(ii)} \end{aligned}\right.$$

Notice from (i) that $(4s + 10)\bar{y} = (2s + 5)\bar{x}$

$$\therefore \quad \boxed{\bar{y} = \frac{\bar{x}}{2}} \qquad \rightarrow \text{That will be a great help! Did you spot it?}$$

Substitute this in (ii)

$$(s + 15)\bar{x} - 18\,.\,\frac{\bar{x}}{2} = 4 + \frac{10}{s}$$

$$\therefore\ (s + 6)\bar{x} = 4 + \frac{10}{s}$$

$$\therefore\ \bar{x} = \frac{4}{s+6} + \frac{10}{s(s+6)} = \frac{4}{s+6} + \frac{5/3}{s} + \frac{5/(-3)}{s+6} = \frac{7}{3}\,.\,\frac{1}{s+6} + \frac{5}{3}\,.\,\frac{1}{s}$$

$$\therefore\ x = \frac{7}{3}\,e^{-6t} + \frac{5}{3}$$

We have seen that $\bar{y} = \frac{\bar{x}}{2}$ $\quad \therefore\ y = \frac{x}{2}$

$$\therefore\ y = \frac{7}{6}\,e^{-6t} + \frac{5}{6}$$

Now on to frame 22.

22

General solutions of equations

One of the strong points about the use of Laplace transforms to solve differential equations is the ability to insert initial conditions very early in the working, so deriving the particular solution required without the necessity of lengthy and tedious substitution at the end.

It is possible, of course, to obtain general solutions to equations if symbols are inserted to represent unknown initial conditions. These will eventually provide the arbitrary constants that appear in any general solution. Let us have an example to show the method.

On then to the next frame.

23

Example 4. Find the general solutions of the equations

$$\begin{cases} \dot{x} + 2y - x = 0 \\ \dot{y} + 2x - y = 0 \end{cases}$$

First express in Laplace transforms:

....

....

24

$$s\bar{x} - x_0 + 2\bar{y} - \bar{x} = 0$$
$$s\bar{y} - y_0 + 2\bar{x} - \bar{y} = 0$$

Now put x_0 = A and y_0 = B in these equations.

Collecting up terms in $\bar{x}$ and $\bar{y}$, the equations become:

....

....

25

$$(s-1)\bar{x} + 2\bar{y} = A$$
$$2\bar{x} + (s-1)\bar{y} = B$$

Now we can eliminate $\bar{y}$ in the usual way and so obtain an expression for $\bar{x}$. You do that.

$\bar{x}$ = ...

26

$$\bar{x} = \frac{As - A - 2B}{(s+1)(s-3)}$$

for we have

$$\begin{cases} (s-1)\bar{x} + 2\bar{y} = A \\ 2\bar{x} + (s-1)\bar{y} = B \end{cases}$$

$$\begin{cases} (s-1)^2\bar{x} + 2(s-1)\bar{y} = A(s-1) \\ 4\bar{x} + 2(s-1)\bar{y} = 2B \end{cases}$$

Subtract $\quad \therefore (s^2 - 2s - 3)\bar{x} = A(s-1) - 2B$

$$\therefore \bar{x} = \frac{As - A - 2B}{(s+1)(s-3)}$$

The next step is, of course, to express this in partial fractions, using the 'cover-up' rule as usual.

$\bar{x} =$ ....

27

$$\bar{x} = \frac{1}{2}\left\{\frac{A+B}{s+1} + \frac{A-B}{s-3}\right\}$$

The constant numerators are getting rather clumsy, so, at this stage, let us denote

$\frac{A+B}{2}$ by P and $\frac{A-B}{2}$ by Q so that

$$\bar{x} = \frac{P}{s+1} + \frac{Q}{s-3}$$

$\therefore x =$ ....

28

$$\boxed{x = \mathrm{P}\,\mathrm{e}^{-t} + \mathrm{Q}\,\mathrm{e}^{3t}}$$

where P and Q are arbitrary constants.

If we were to start afresh to find the solution for y, we should arrive at a similar solution with two more arbitrary constants, apparently giving us a total of *four* arbitrary constants for the complete solution. But two first order equations can have a total of only *two* arbitrary constants between them, so the new pair must be related to those we have already obtained. We therefore use the result for x to determine the solution for y.

Now the first of the two given equations was

$$\dot{x} + 2y - x = 0$$

$\therefore$ Knowing x and $\dot{x}$ (which we can soon get by differentiating our expression for x), we can find y.

$$x = \mathrm{P}\,\mathrm{e}^{-t} + \mathrm{Q}\,\mathrm{e}^{3t}$$

$$\therefore\ \dot{x} = \ldots\ldots\ldots\ldots\ldots\ldots\ldots\ldots$$

29

$$\boxed{\dot{x} = -\mathrm{P}\,\mathrm{e}^{-t} + 3\mathrm{Q}\,\mathrm{e}^{3t}}$$

So $$\dot{x} + 2y - x = 0 \qquad \therefore\ 2y = x - \dot{x}$$

$$\therefore\ 2y = (\mathrm{P}\,\mathrm{e}^{-t} + \mathrm{Q}\,\mathrm{e}^{3t}) - (-\mathrm{P}\,\mathrm{e}^{-t} + 3\mathrm{Q}\,\mathrm{e}^{3t})$$

and, collecting up terms, this finally gives

$$y = \ldots\ldots\ldots\ldots\ldots\ldots\ldots\ldots\ldots\ldots\ldots\ldots$$

30

$$\boxed{y = \mathrm{P}\,\mathrm{e}^{-t} - \mathrm{Q}\,\mathrm{e}^{3t}}$$

So the required solutions are:

$$\underline{x = \mathrm{P}\,\mathrm{e}^{-t} + \mathrm{Q}\,\mathrm{e}^{3t}}; \qquad \underline{y = \mathrm{P}\,\mathrm{e}^{-t} - \mathrm{Q}\,\mathrm{e}^{3t}}$$

On now to frame 31.

31

Second order simultaneous differential equations

So far we have been concerned with pairs of first order differential equations. We deal with second order equations in very much the same way.

Example 5. Solve:

$$\left.\begin{aligned}\ddot{x} + 3x - 2y &= 0\\ \ddot{x} + \ddot{y} - 3x + 5y &= 0\end{aligned}\right\}\quad \text{given that at } t = 0,\ x = y = 0,\ \dot{x} = 1,\ \dot{y} = 3.$$

First we re-write these in terms of the transforms.

Remembering that

$$\mathcal{L}\{\ddot{x}\} = s^2\bar{x} - sx_0 - x_1$$

and

$$\mathcal{L}\{\ddot{y}\} = s^2\bar{y} - sy_0 - y_1$$

the equations become

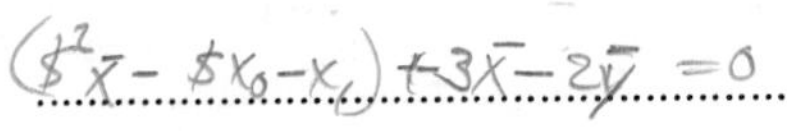

....

....

32

$$\boxed{\begin{aligned}(s^2\bar{x} - sx_0 - x_1) + 3\bar{x} - 2\bar{y} = 0\\ (s^2\bar{x} - sx_0 - x_1) + (s^2\bar{y} - sy_0 - y_1) - 3\bar{x} + 5\bar{y} = 0\end{aligned}}$$

In this case,

$$x_0 = y_0 = 0$$

$$x_1 = 1; \qquad y_1 = 3$$

So now insert the initial conditions and, as usual, collect up terms in $\bar{x}$ and $\bar{y}$

....

....

33

$$(s^2 + 3)\bar{x} - 2\bar{y} = 1$$
$$(s^2 - 3)\bar{x} + (s^2 + 5)\bar{y} = 4$$

For we have:

$$\left.\begin{aligned} s^2\bar{x} - 1 + 3\bar{x} - 2\bar{y} &= 0 \\ s^2\bar{x} - 1 + s^2\bar{y} - 3 - 3\bar{x} + 5\bar{y} &= 0 \end{aligned}\right\}$$

$$\therefore \quad \begin{cases} (s^2 + 3)\bar{x} - 2\bar{y} = 1 \\ (s^2 - 3)\bar{x} + (s^2 + 5)\bar{y} = 4 \end{cases}$$

Now eliminate $\bar{y}$ to obtain an expression for $\bar{x}$.

$\bar{x} =$ ....

34

$$\bar{x} = \frac{s^2 + 13}{(s^2 + 1)(s^2 + 9)}$$

Here are the steps:

$$\begin{cases} (s^2 + 3)(s^2 + 5)\bar{x} - 2(s^2 + 5)\bar{y} = s^2 + 5 \\ 2(s^2 - 3)\bar{x} + 2(s^2 + 5)\bar{y} = 8 \end{cases}$$

Adding these, we get

$$(s^4 + 8s^2 + 15 + 2s^2 - 6)\bar{x} = s^2 + 13$$
$$(s^4 + 10s^2 + 9)\bar{x} = s^2 + 13$$
$$(s^2 + 1)(s^2 + 9)\bar{x} = s^2 + 13$$
$$\therefore \ \bar{x} = \frac{s^2 + 13}{(s^2 + 1)(s^2 + 9)}$$

Since there is no term in s, we can treat s^2 as a single variable and express $\bar{x}$ in partial fractions by using the 'cover-up' rule.

$\bar{x} =$ ....

35

$$\boxed{\bar{x} = \frac{3}{2}\cdot\frac{1}{s^2+1} - \frac{1}{2}\cdot\frac{1}{s^2+9}}$$

for
$$\bar{x} = \frac{s^2+13}{(s^2+1)(s^2+9)}$$
$$= \frac{12/8}{s^2+1} + \frac{4/(-8)}{s^2+9}$$
$$= \frac{3}{2}\cdot\frac{1}{s^2+1} - \frac{1}{2}\cdot\frac{1}{s^2+9}$$

∴ Taking inverse transforms

$x =$ ....

36

$$\boxed{x = \frac{3}{2}\sin t - \frac{1}{6}\sin 3t}$$

To find y we could go back and eliminate $\bar{x}$, and then proceed as before. However, keeping our eye open for possible useful variations, in this case, there is another route.

The first of the given equations is

$$\ddot{x} + 3x - 2y = 0$$

We can therefore easily obtain y by differentiating the solution for x twice and substituting for $\ddot{x}$ and x.

If we do that, we obtain:

$y =$ ....

37

$$\boxed{y = \frac{3}{2}\sin t + \frac{1}{2}\sin 3t}$$

for

$$x = \frac{3}{2}\sin t - \frac{1}{6}\sin 3t$$

$$\therefore \dot{x} = \frac{3}{2}\cos t - \frac{1}{2}\cos 3t$$

$$\therefore \ddot{x} = -\frac{3}{2}\sin t + \frac{3}{2}\sin 3t$$

$$\therefore 2y = \ddot{x} + 3x$$

$$= -\frac{3}{2}\sin t + \frac{3}{2}\sin 3t + \frac{9}{2}\sin t - \frac{1}{2}\sin 3t = 3\sin t + \sin 3t$$

$$\therefore \underline{y = \frac{3}{2}\sin t + \frac{1}{2}\sin 3t}$$

Collecting our two results, we have

$$\underline{x = \frac{3}{2}\sin t - \frac{1}{6}\sin 3t}; \qquad \underline{y = \frac{3}{2}\sin t + \frac{1}{2}\sin 3t}$$

On to the next frame.

38

In some practical problems, we need to obtain the solution for only one of the variables concerned. This means, of course, that we eliminate the second variable, in the way we have been doing, and give no further attention to it.

Here is a further example.

Example 6. Find an expression for y given that

$$\ddot{x} - \ddot{y} + x - y = 5\,e^{2t}$$

$$2\dot{x} - \dot{y} + y = 0$$

and that at $t = 0, x = 1, y = 2, \dot{x} = 0$.

As usual, first express the equations in terms of the transforms.

....

....

39

$$(s^2\bar{x} - sx_0 - x_1) - (s^2\bar{y} - sy_0 - y_1) + \bar{x} - \bar{y} = \frac{5}{s-2}$$

$$2(s\bar{x} - x_0) - (s\bar{y} - y_0) + \bar{y} = 0$$

Now we insert the initial conditions.

In this case, we are told that $x_0 = 1; y_0 = 2; x_1 = 0$.

So we get

....

....

40

$$s^2\bar{x} - s - s^2\bar{y} + 2s + y_1 + \bar{x} - \bar{y} = \frac{5}{s-2}$$

$$2s\bar{x} - 2 - s\bar{y} + 2 + \bar{y} = 0$$

But what about the value of y_1? It looks as though it has been omitted by mistake.

It is, of course, highly likely that we are given sufficient information for our needs if only we can recognize it. So, although we are not specifically told the value of y_1, can we in fact use the information we are given to obtain the value of y_1?

Refer back to the original equations in the question and see what can be found. After all that,

$$y_1 = \text{..}$$

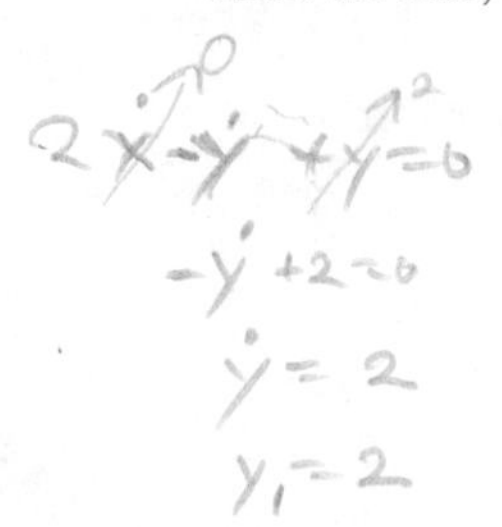

41

$$\boxed{y_1 = 2}$$

since the second equation is $2\dot{x} - \dot{y} + y = 0$ and we are given that, at $t = 0, x = 1$, $y = 2, \dot{x} = 0$ i.e. $x_0 = 1, y_0 = 2, x_1 = 0$

$$\therefore\ 0 - \dot{y} + 2 = 0 \qquad \therefore\ \dot{y} = 2 \qquad \therefore\ y_1 = 2$$

So now we can put this into the transformed equations and collect up terms.

....

....

42

$$\boxed{\begin{array}{c}(s^2 + 1)\bar{x} - (s^2 + 1)\bar{y} = \dfrac{5}{s-2} - s - 2 \\[2ex] 2s\bar{x} - (s-1)\bar{y} = 0\end{array}}$$

Eliminating x, we obtain

$$\begin{cases} 2s(s^2 + 1)\bar{x} - 2s(s^2 + 1)\bar{y} = \dfrac{10s}{s-2} - 2s^2 - 4s & \ldots \quad \text{(i)} \\[2ex] 2s(s^2 + 1)\bar{x} - (s-1)(s^2 + 1)\bar{y} = 0 & \ldots \quad \text{(ii)} \end{cases}$$

$$\text{(ii)} - \text{(i)} \qquad \therefore\ (s + 1)(s^2 + 1)\bar{y} = 2s^2 + 4s - \frac{10s}{s-2}$$

This time it will pay to combine the right-hand-side terms into one fraction.

i.e. $$\frac{2s(s+2)(s-2) - 10s}{s-2} = \frac{2s(s^2 - 4 - 5)}{s-2} = \frac{2s(s^2 - 9)}{s-2}$$

$$\therefore\ (s + 1)(s^2 + 1)\bar{y} = \frac{2s(s^2 - 9)}{s-2}$$

$\therefore\ \bar{y} =$ ....

43

$$\boxed{\bar{y} = \frac{2s(s^2 - 9)}{(s+1)(s-2)(s^2+1)}}$$

Now for partial fractions. By the usual method, we have:

$$\bar{y} = \frac{(-2)(-8)/(-3)(2)}{s+1} + \frac{(4)(-5)/(3)(5)}{s-2} + \frac{?}{s^2+1}$$

$$= -\frac{8}{3}\cdot\frac{1}{s+1} - \frac{4}{3}\cdot\frac{1}{s-2} + \frac{?}{s^2+1}$$

The numerator of the third term will be of the form (As + B) and to find A and B we combine the first two terms.

$$\bar{y} = -\frac{4}{3}\left(\frac{2}{s+1} + \frac{1}{s-2}\right) + \frac{\mathrm{A}s + \mathrm{B}}{s^2+1}$$

$$= -\frac{4}{3}\left\{\frac{3s-3}{(s+1)(s-2)}\right\} + \frac{\mathrm{A}s + \mathrm{B}}{s^2+1}$$

$$= \frac{-4(s-1)}{s^2 - s - 2} + \frac{\mathrm{A}s + \mathrm{B}}{s^2+1} \equiv \frac{2s(s^2-9)}{(s+1)(s-2)(s^2+1)}$$

Equating coefficients of s^3 and also the constant terms, this gives

A = ...; B = ...

44

$$\boxed{A = 6; \qquad B = 2}$$

since (i) coefficient of s^3: $\quad -4 + A = 2 \qquad \therefore A = 6$

(ii) const. terms: $\quad 4 - 2B = 0 \qquad \therefore B = 2$

$$\therefore \; \bar{y} = -\frac{8}{3} \cdot \frac{1}{s+1} - \frac{4}{3} \cdot \frac{1}{s-2} + \frac{6s+2}{s^2+1}$$

$\therefore \; y =$ ....

45

$$\boxed{y = -\frac{8}{3}\, e^{-t} - \frac{4}{3} e^{2t} + 6 \cos t + 2 \sin t}$$

That is the finish of that problem since we were asked to find an expression for y only.

Finally, here is one further example for you to work through entirely on your own.

Example 7. Find an expression for y, given that

$$\ddot{y} - 5y - 5x = 0$$

$$\ddot{x} + x + y = 0$$

and that at $t = 0, x = y = 0; \dot{x} = 0; \dot{y} = 2$.

Complete the solution as required

$y =$ ....

and then check your working with the next frame.

46

$$y = \frac{5}{4}\sinh 2t - \frac{t}{2}$$

Here is the working:

$$\left.\begin{aligned} \ddot{y} - 5y - 5x &= 0 \\ \ddot{x} + x + y &= 0 \end{aligned}\right\} \quad t = 0; \quad x = y = 0; \quad \dot{x} = 0; \quad \dot{y} = 2$$

$$\left.\begin{aligned} (s^2\bar{y} - sy_0 - y_1) - 5\bar{y} - 5\bar{x} &= 0 \\ (s^2\bar{x} - sx_0 - x_1) + \bar{x} + \bar{y} &= 0 \end{aligned}\right\} \quad x_0 = y_0 = x_1 = 0; y_1 = 2.$$

$$\therefore \begin{cases} s^2\bar{y} - 2 - 5\bar{y} - 5\bar{x} = 0 \\ s^2\bar{x} + \bar{x} + \bar{y} = 0 \end{cases}$$

$$\therefore \begin{cases} (s^2 - 5)\bar{y} - 5\bar{x} = 2 \\ \bar{y} + (s^2 + 1)\bar{x} = 0 \end{cases}$$

To find $\bar{y}$, eliminate $\bar{x}$

$$\therefore \begin{cases} (s^2 + 1)(s^2 - 5)\bar{y} - 5(s^2 + 1)\bar{x} = 2(s^2 + 1) \\ 5\bar{y} + 5(s^2 + 1)\bar{x} = 0 \end{cases}$$

Add: $\therefore (s^4 - 4s^2 - 5 + 5)\bar{y} = 2(s^2 + 1)$

$$s^2(s^2 - 4)\bar{y} = 2(s^2 + 1)$$

$$\therefore \bar{y} = \frac{2(s^2 + 1)}{s^2(s^2 - 4)}$$

$$= \frac{2/(-4)}{s^2} + \frac{10/4}{s^2 - 4}$$

$$= -\frac{1}{2} \cdot \frac{1}{s^2} + \frac{5}{2} \cdot \frac{1}{s^2 - 4}$$

$$\therefore y = \frac{5}{4}\sinh 2t - \frac{t}{2}$$

Of course, since $\sinh 2t = \dfrac{e^{2t} - e^{-2t}}{2}$, you may have the result in its exponential form. In that case

$$y = \frac{5}{8}(e^{2t} - e^{-2t}) - \frac{t}{2}$$

That was all we were asked to do. But, just by way of revision, go ahead now and find the solution for x – and incidentally, do not forget any possible short cuts. When you have finished, check with the next frame.

47

$$\boxed{x = \frac{t}{2} - \frac{1}{4}\sinh 2t}$$

The previous result gave us $y = \frac{5}{4}\sinh 2t - \frac{t}{2}$ and the first of the given equations is

$$\ddot{y} - 5y - 5x = 0$$

$\therefore$ Substitute for $\ddot{y}$ and y and so obtain an expression for x.

$$y = \frac{5}{4}\sinh 2t - \frac{t}{2}$$

$$\therefore\ \dot{y} = \frac{5}{2}\cosh 2t - \frac{1}{2}$$

$$\therefore\ \ddot{y} = 5\sinh 2t$$

$$\therefore\ 5x = 5\sinh 2t - \frac{25}{4}\sinh 2t + \frac{5t}{2}$$

$$\therefore\ x = \sinh 2t - \frac{5}{4}\sinh 2t + \frac{t}{2}$$

$$\therefore\ x = \frac{t}{2} - \frac{1}{4}\sinh 2t$$

So the complete pair of solutions is

$$x = \frac{t}{2} - \frac{1}{4}\sinh 2t; \qquad y = \frac{5}{4}\sinh 2t - \frac{t}{2}$$

And that brings us almost to the end of this programme. Once again, the method depends on a sound knowledge of partial fractions and care with simple algebraic processes.

Before working through the Text Exercise, revise any part of the programme about which you do not feel perfectly clear. There is no rush. When you are ready, turn on to the next frame and complete the exercise.

48

Work through all the problems below. They are quite straightforward and contain no tricks, so will not cause difficulty. Take your time.

Test Exercise–IV

1. Solve the equations

$$\dot{y} + 2x = e^{-t}$$

$$\dot{x} - 2y = e^{t}$$

given that at $t = 0, x = 0, y = 0$.

2. Solve the equations

$$2\dot{x} - 6x + 3y = 0$$

$$3\dot{y} - 3y - 2x = 0$$

given that at $t = 0, x = 3, y = 1$.

3. Find the general solution of the equations

$$\dot{x} - 2x + y = 0$$

$$\dot{y} + 2x - 3y = 0$$

4. Find an expression for x, given that

$$\ddot{x} + 3\dot{x} - 2x + \dot{y} - 3y = 2\,e^{2t}$$

$$2\dot{x} - x + \dot{y} - 2y = 0$$

and that at $t = 0, x = \dot{x} = 0, y = 4$.

5. If $\ddot{x} - x + 5\dot{y} = t$

$$\ddot{y} - 4y - 2\dot{x} = -2$$

and at $t = 0, x = 0, \dot{x} = 0, y = 1, \dot{y} = 0$, find an expression for y in terms of t.

Further Problems IV

Solve the following pairs of simultaneous equations with the stated initial conditions:

1. $\left.\begin{array}{l}\dot{x}+5x+y=\mathrm{e}^{-t}\\ \dot{y}-x+3y=\mathrm{e}^{-2t}\end{array}\right\}\quad t=0,\ x=\tfrac{1}{2},\ y=\tfrac{2}{3}.$

2. $\left.\begin{array}{l}\dot{y}+2y+2x=4\\ \dot{x}+y+3x=10\,\mathrm{e}^{-t}\end{array}\right\}\quad t=0,\ y=0,\ \dot{x}=0.$

3. $\left.\begin{array}{l}\dot{x}+\dot{y}=t\\ \ddot{x}-y=\mathrm{e}^{-t}\end{array}\right\}\quad t=0,\ x=3,\ \dot{x}=-2,\ y=0.$

4. $\left.\begin{array}{l}\ddot{x}+5\dot{y}-4x=3\sin 2t\\ \ddot{y}-5\dot{x}-4y=0\end{array}\right\}\quad t=0,\ x=y=\dot{y}=0,\ \dot{x}=1.$

5. $\left.\begin{array}{l}\ddot{x}-2x-3y=\mathrm{e}^{2t}\\ \ddot{y}+2y+x=0\end{array}\right\}\quad t=0,\ x=y=1,\ \dot{x}=\dot{y}=0$

6. $\left.\begin{array}{l}\dot{x}+\dot{y}+2x+y=\mathrm{e}^{-3t}\\ \dot{y}+5x+3y=5\,\mathrm{e}^{-2t}\end{array}\right\}\quad t=0,\ x=-1,\ y=4.$

7. $\left.\begin{array}{l}2\dot{x}-2\dot{y}-9y=\mathrm{e}^{-2t}\\ 2\dot{x}+4\dot{y}+4x-37y=0\end{array}\right\}\quad t=0,\ x=0,\ y=\tfrac{1}{4}.$

8. $\left.\begin{array}{l}\mathrm{L}\dot{x}+\mathrm{R}x+\mathrm{R}(x-y)=\mathrm{E}\\ \mathrm{L}\dot{y}+\mathrm{R}y-\mathrm{R}(x-y)=0\end{array}\right\}\quad t=0,\ x=y=0\ (\mathrm{L},\mathrm{R},\mathrm{E}\text{ constants})$

9. $\left.\begin{array}{l}2\dot{x}+2x+3\dot{y}+6y=56\,\mathrm{e}^{t}-3\,\mathrm{e}^{-t}\\ \dot{x}-2x-\dot{y}-3y=-21\,\mathrm{e}^{t}-7\,\mathrm{e}^{-t}\end{array}\right\}\quad \text{At } t=0,\ x=8,\ y=3.$

10. $\left.\begin{array}{l}\ddot{x}+x=\dot{y}\\ 4\dot{x}+2x=\dot{y}+2y\end{array}\right\}\quad t=0,\ x=0,\ y=1,\ \dot{x}=2.$

11. $\left.\begin{array}{l}\dot{x}+2\dot{y}-3x=-\mathrm{e}^{-2t}\\ 2\dot{x}-3x-4y=3\,\mathrm{e}^{-t}-3\mathrm{e}^{-2t}\end{array}\right\}\quad t=0,\ x=y=0.$

12. $\dot{y} - \dot{x} - 2y + 2x = \sin t$, $\ddot{y} + 2\dot{x} + y = 0$ Zero initial conditions.

13. A unit e.m.f. is applied at $t = 0$, to the network shown.

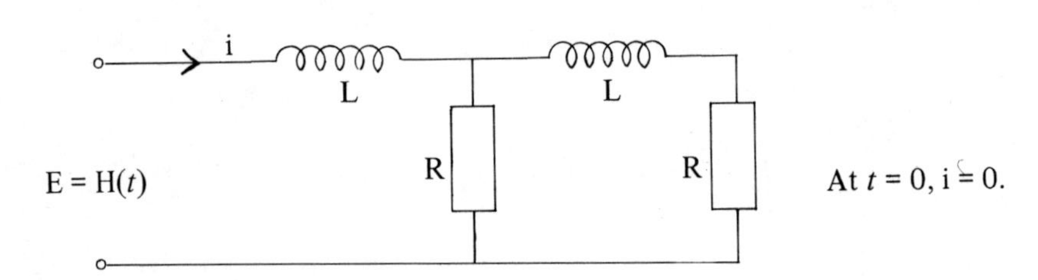

Show that, at time t,

$$i = \frac{2}{R}\left\{1 - e^{\frac{-3Rt}{2L}}\left(\cosh\frac{\sqrt{5}.R_t}{2L} + \frac{2}{\sqrt{5}}\sinh\frac{\sqrt{5}.R}{2L}t\right)\right\}$$

14. Given that $\ddot{y} - \dot{x} + 2x = 10 \sin 2t$

$$\dot{y} + 2y + x = 0$$

and that $y = x = 0$ when $t = 0$, find an expression for x in terms of t.

15. Solve $\left.\begin{array}{l} 5a\ddot{\theta} + 12a\ddot{\phi} + 6g\theta = 0 \\ 5a\ddot{\theta} + 16a\ddot{\phi} + 6g\phi = 0 \end{array}\right\}$

given that when $t = 0$, $\theta = \dfrac{7a}{4}$, $\phi = a$, $\dot{\theta} = \dot{\phi} = 0$.

16. If an electron is projected into a uniform magnetic field perpendicular to its direction of motion, its path is given by

$$m\ddot{y} = -\frac{He}{c}\dot{x}; \qquad m\ddot{x} = \frac{He}{c}\dot{y}$$

with $x = 0$, $\dot{x} = u$, $y = \dot{y} = 0$ at $t = 0$. H, m, e, c are constants. Find x and y in terms of t.

17. Find an expression for y in terms of t, given that

$$\ddot{x} + 8x + 2y = 24 \cos 4t$$

$$\ddot{y} + 2x + 5y = 0$$

and that at $t = 0$, $x = 0$, $y = 0$, $\dot{x} = 1$, $\dot{y} = 2$.

18. The equations of motion of two masses attached to a horizontal elastic string and with separate displacements, x and y, are given by

$$\ddot{x} + 2x - y = 0$$

$$\ddot{y} + 2y - x = 0$$

Determine x and y in terms of t, if the system is released from rest at $t = 0$ with $x = a$ and $y = a/2$.

19. Solve $\ddot{x} + 2\dot{y} + x = a \sin t$

$$\ddot{y} + 2\dot{x} + y = a \cos t$$

if $x = x = \dot{x} = \dot{y} = 0$, at $t = 0$.

20. Currents i_1 and i_2 in a network are related by the following equations

$$4\frac{di_1}{dt} - 2\frac{di_2}{dt} + 10\,i_1 - 5\,i_2 = 0$$

$$\frac{di_2}{dt} + 5\,i_1 + 15\,i_2 = 35$$

At $t = 0$, $i_1 = 2$, $i_2 = 4$. Solve for i_1 and i_2.

Programme 5

HEAVISIDE UNIT STEP FUNCTION

1

Introduction

So far, our work on Laplace transforms has been concerned with continuous functions of t. In practical applications, we are often required to deal with situations that change abruptly at specified values of t. It is convenient, therefore, to have a function which will 'switch-on' or 'switch-off' a given term at certain stated values of t, and this facility is provided by the *Heaviside unit step function.*

Heaviside Unit Step Function

Consider a function which has zero value for all values of t up to $t = c$ and a unit value for $t = c$ and all values $t > c$.

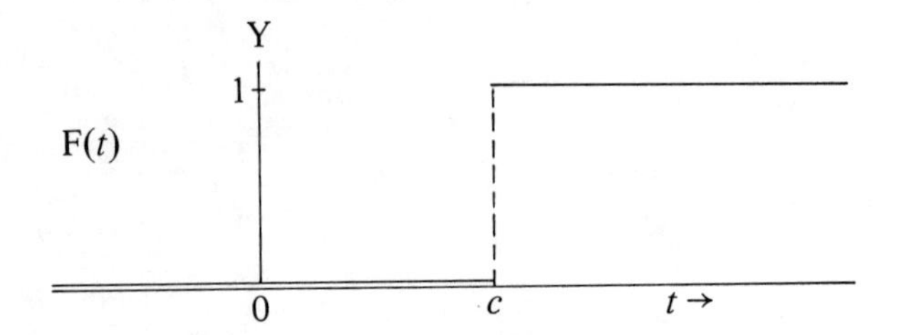

$$F(t) = 0 \text{ for } t < c$$
$$= 1 \text{ for } t \geqslant c$$

This is the Heaviside unit step function and is denoted by

$$F(t) = H(t - c)$$

where the constant c indicates the value of t at which the function changes from a value 0 to a value 1.

So we denote the function:

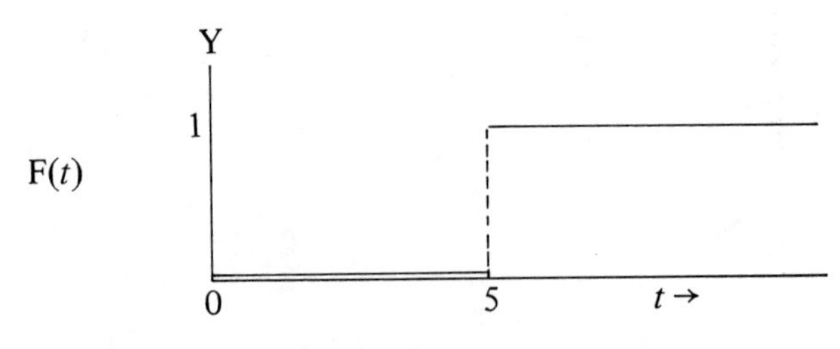

as $F(t) =$ ....

2

$$\boxed{F(t) = H(t-5)}$$

Remember that $H(t-5)$ always has one of two values, ... or ....

3

$$\boxed{0 \text{ or } 1}$$

For $t < 5$, $H(t-5)$ has a value ....

$t \geqslant 5$, $H(t-5)$ has a value ....

4

$$\boxed{t < 5,\ H(t-5) = 0; \qquad t \geqslant 5,\ H(t-5) = 1}$$

$H(t-c)$ is thus a two-valued function

$$H(t-c) = 0 \text{ for } t < c$$

$$H(t-c) = 1 \text{ for } t \geqslant c.$$

If the step occurs at the origin

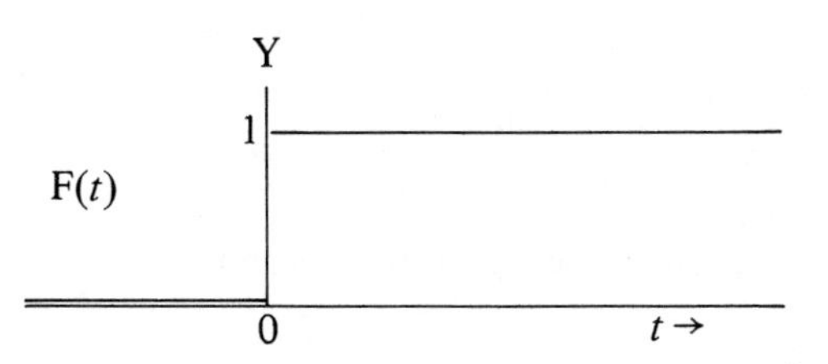

then, in this case,

$F(t) =$ ....

5

$$\boxed{F(t) = H(t)}$$

Since $c = 0$ $\quad \therefore \; F(t) = H(t - 0) = H(t)$.

The graph of $F(t) = e^{-t}$ is, of course, as shown

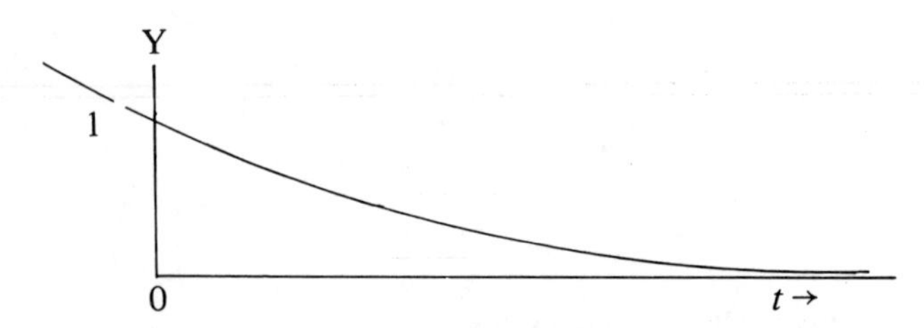

Remembering that $H(t - c) = 0$ for $t < c$

$= 1$ for $t \geqslant c$

sketch the graph of

$$F(t) = H(t - c)\, e^{-t}$$

When you have sketched it, move on to frame 6.

6

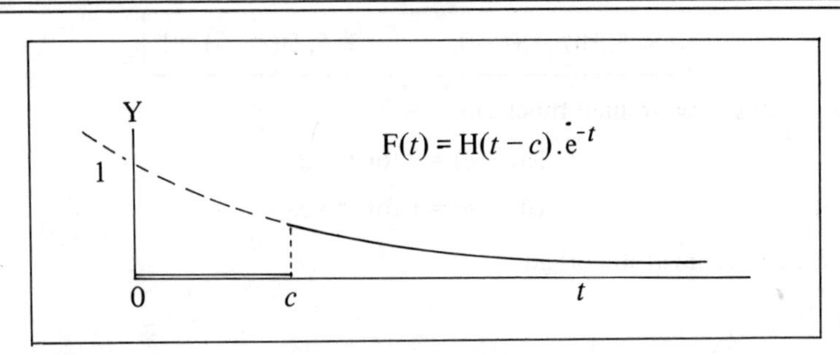

For $t < c$, $H(t - c) = 0$ $\quad \therefore \; H(t - c).e^{-t} = 0.e^{-t} = 0$.

For $t \geqslant c$, $H(t - c) = 1$ $\quad \therefore \; H(t - c).e^{-t} = 1.e^{-t} = e^{-t}$.

The function $H(t - c)$ thus suppresses the function e^{-t} prior to $t = c$ and 'switches-on' the function at $t = c$.

In the same way, sketch

(i) the graph of $F(t) = \sin t$ for one complete cycle, $0 < t < 2\pi$,

then (ii) the graph of $F(t) = H(t - \pi/3).\sin t$ for the same interval.

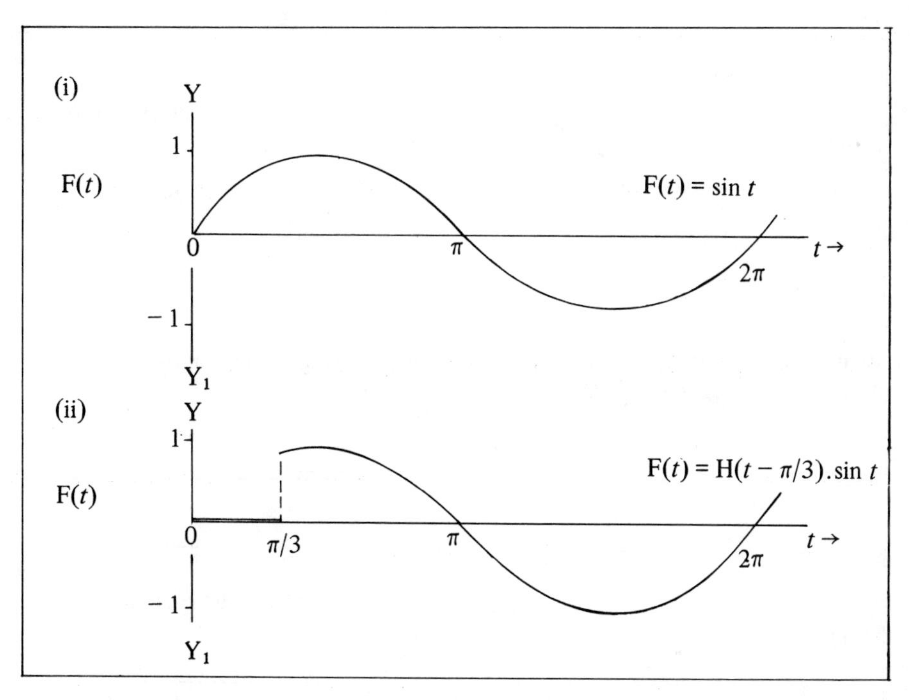

In (ii), as before, $H(t - \pi/3) = 0$ for all values of t before $t = \pi/3$, i.e. $F(t) = 0$.

$H(t - \pi/3) = 1$ at $t = \pi/3$ and for all values of $t > \pi/3$.

$\therefore$ From $t = \pi/3$, $F(t) = \sin t$.

$\therefore$ We have the graph of $F(t) = \sin t$, but suppressed up to $t = \pi/3$.

On to frame 8.

8

Let us now consider the graph of $F(t) = H(t - c) . \sin (t - c)$.
First sketch the graph of $F(t) = \sin (t - c)$.
For convenience, take $c = \pi/3$.

9

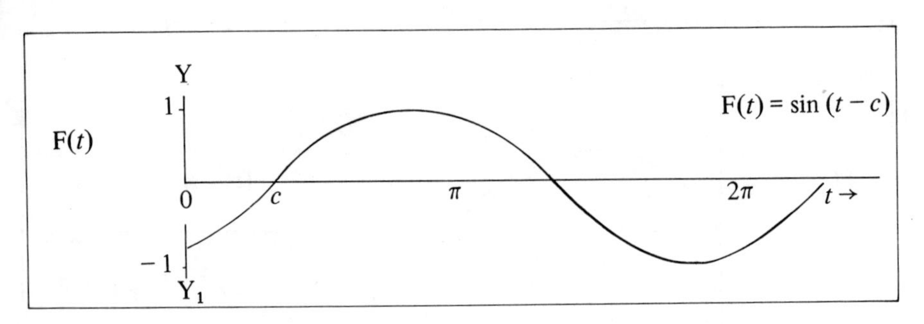

Remembering that $H(t-c)$ has the effect of suppressing a function for $t < c$, now sketch the graph of $F(t) = H(t-c).\sin(t-c)$.

10

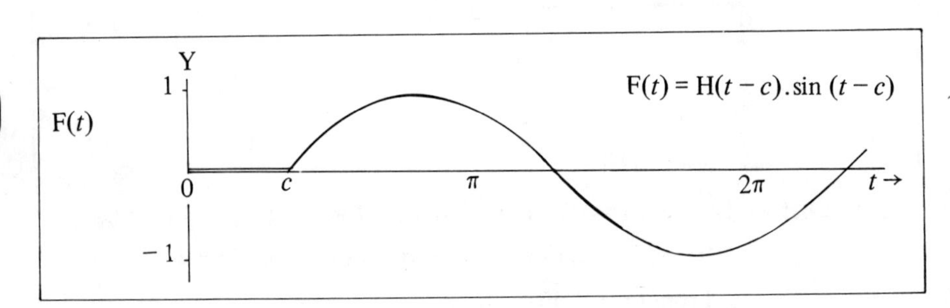

Thus the graph of $F(t) = H(t-c) \, . \, \sin(t-c)$ is the graph of $F(t) = \sin t \; (t > 0)$, *shifted along the t-axis through an interval c units.*

At $t = c$, $\sin(t-c) = \sin 0 = 0$, i.e. the value of $\sin t$ at $t = 0$.

....

So, by way of revision, sketch the following three graphs separately and under each other to show their differences. Arrange the graphs to show one cycle and let $c = \pi/3$ as before.

(i) $F(t) = \cos t$

(ii) $F(t) = H(t-c).\cos t$

(iii) $F(t) = H(t-c).\cos(t-c)$.

11

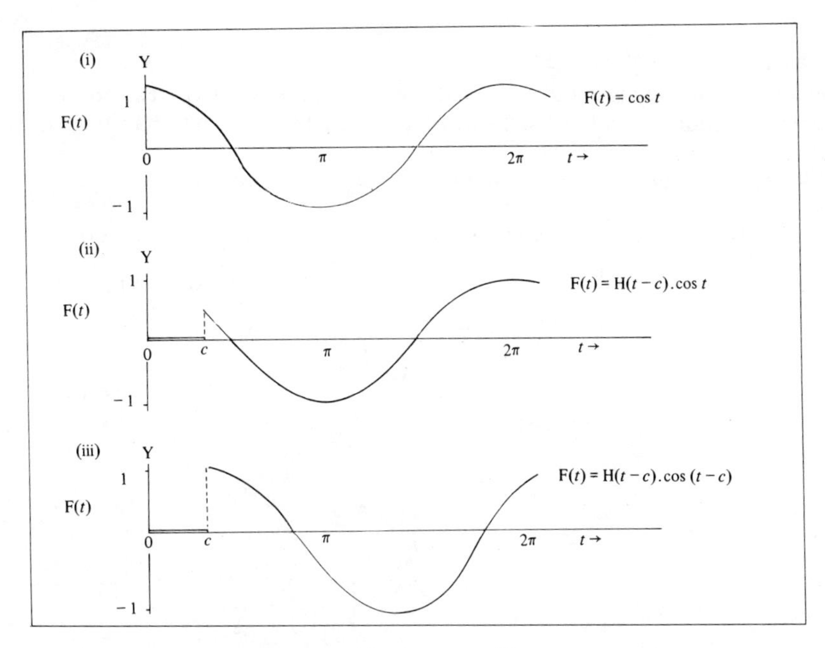

In (ii), $F(t) = \cos t$ is suppressed before $t = \pi/3$
In (iii), $F(t) = \cos t$ $(t > 0)$ is shifted $\pi/3$ units to the right along the t-axis.

Summary: In general, then, we have:

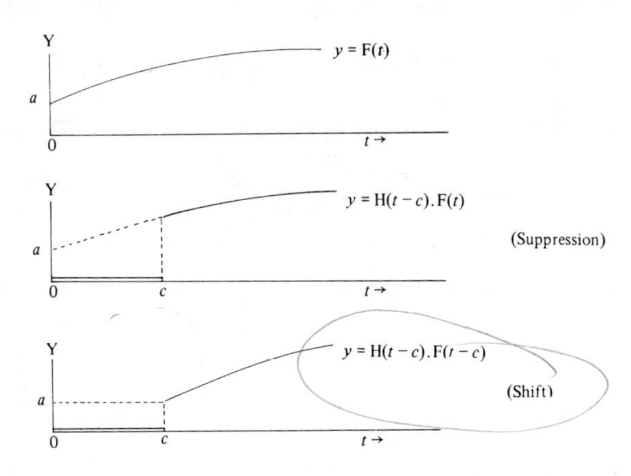

Now on to the next frame.

12

Laplace transform of H($t-c$)

Since a function involving the Heaviside unit step function may well appear as a forcing function of a differential equation, we need to be able to transform it with the rest of the terms.

We simply apply the basic definition of a Laplace transform.

$$\mathcal{L}\{\mathrm{H}(t-c)\} = \int_0^\infty \mathrm{e}^{-st}\,\mathrm{H}(t-c)\,dt \qquad \begin{array}{ll} 0<t<c, & \mathrm{H}(t-c)=0 \\ t\geqslant c, & \mathrm{H}(t-c)=1 \end{array}$$

$$\therefore\ \mathrm{e}^{-st}.\mathrm{H}(t-c) = 0 \text{ between } t=0 \text{ and } t=c,$$

$$= \mathrm{e}^{-st} \text{ for } t \geqslant c.$$

$$\therefore\ \mathcal{L}\{\mathrm{H}(t-c)\} = \int_0^\infty \mathrm{e}^{-st}\,\mathrm{H}(t-c)\,dt$$

$$= \int_c^\infty \mathrm{e}^{-st}\,dt = \left[\frac{\mathrm{e}^{-st}}{-s}\right]_c^\infty$$

$$= \{0\} - \left\{\frac{\mathrm{e}^{-cs}}{-s}\right\} = \frac{\mathrm{e}^{-cs}}{s}$$

$$\therefore\ \mathcal{L}\{\mathrm{H}(t-c)\} = \frac{\mathrm{e}^{-cs}}{s}$$

This is important: make a note of this result in your record book.

The Laplace transform of the unit step function *at the origin* will therefore be

....

13

$$\boxed{\mathcal{L}\{\mathrm{H}(t)\} = \frac{1}{s}}$$

Since $\mathcal{L}\{\mathrm{H}(t-c)\} = \dfrac{\mathrm{e}^{-cs}}{s}$ and, for the unit step function at the origin, $c = 0$.

$$\therefore\ \mathcal{L}\{\mathrm{H}(t)\} = \frac{\mathrm{e}^{0}}{s} = \frac{1}{s}$$

So we have:

$$\mathcal{L}\{\mathrm{H}(t-c)\} = \frac{\mathrm{e}^{-cs}}{s}$$

and

$$\mathcal{L}\{\mathrm{H}(t)\} = \frac{1}{s}$$

Laplace transform of $\mathrm{H}(t-c).\mathrm{F}(t-c)$

Applying the definition of a Laplace transform,

$$\mathcal{L}\{\mathrm{H}(t-c).\mathrm{F}(t-c)\} = \int_0^\infty \mathrm{e}^{-st}\,\mathrm{H}(t-c)\,\mathrm{F}(t-c)\,dt$$

Since $\mathrm{H}(t-c) = 0$ for $0 < t < c$

$= 1$ for $t \geqslant c$

$$\therefore\ \mathcal{L}\{\mathrm{H}(t-c).\mathrm{F}(t-c)\} = \int_c^\infty \mathrm{e}^{-st}.\mathrm{F}(t-c)\,dt$$

Let us now put $t - c = v \qquad \therefore\ t = c + v \qquad \therefore\ dt \equiv dv.$

Also when $t = c$, $v = 0$ and when $t = \infty$, $v = \infty$.

$$\therefore\ \mathcal{L}\{\mathrm{H}(t-c).\mathrm{F}(t-c)\} = \int_0^\infty \mathrm{e}^{-s(c+v)}.\mathrm{F}(v)\,dv$$

$$= \mathrm{e}^{-cs}\int_0^\infty \mathrm{e}^{-sv}.\mathrm{F}(v)\,dv.$$

Since the integration is between limits, the result will depend on those limits and not on the particular symbol chosen for the variable.

e.g. $\int_0^\infty \mathrm{e}^{-sv}\,\mathrm{F}(v)\,dv$ has exactly the same value as $\int_0^\infty \mathrm{e}^{-st}\,\mathrm{F}(t)\,dt$

i.e. $\int_0^\infty \mathrm{e}^{-sv}.\mathrm{F}(v)\,dv = \int_0^\infty \mathrm{e}^{-st}.\mathrm{F}(t)\,dt = \mathcal{L}\{\mathrm{F}(t)\} = f(s).$

$\therefore\ \mathcal{L}\{\mathrm{H}(t-c).\mathrm{F}(t-c)\} = \mathrm{e}^{-cs}.\mathcal{L}\{\mathrm{F}(t)\} = \mathrm{e}^{-cs}.f(s)$

$$\underline{\mathcal{L}\{\mathrm{H}(t-c).\mathrm{F}(t-c)\} = \mathrm{e}^{-cs}.f(s)} \quad \text{where} \quad \underline{f(s) = \mathcal{L}\{\mathrm{F}(t)\}}.$$

Make a note of this important result.

15

$$\mathcal{L}\{\mathrm{H}(t-c).\mathrm{F}(t-c)\} = \mathrm{e}^{-cs} f(s) \text{ where } f(s) = \mathcal{L}\{\mathrm{F}(t)\}$$

Example 1. $\mathcal{L}\{\mathrm{H}(t-3).(t-3)^2\} = \mathrm{e}^{-3s}.f(s) \quad \text{where} \quad f(s) = \mathcal{L}\{t^2\} = \dfrac{2!}{s^3}$

$$= \frac{\mathrm{e}^{-3s}.2}{s^3} = \frac{2\,\mathrm{e}^{-3s}}{s^3}$$

Example 2. $\mathcal{L}\{\mathrm{H}(t-2).\sin(t-2)\} = \mathrm{e}^{-2s}.f(s)$

where $f(s) =$ ....

16

$$f(s) = \mathcal{L}\{\sin t\} = \frac{1}{s^2+1}$$

$\therefore\ \mathcal{L}\{\mathrm{H}(t-2)\sin(t-2)\} =$ ....

17

$$\frac{\mathrm{e}^{-2s}}{s^2+1}$$

For: $\mathcal{L}\{\mathrm{H}(t-2)\sin(t-2)\} = \mathrm{e}^{-2s}.f(s) \quad \text{where} \quad f(s) = \mathcal{L}\{\sin t\} = \dfrac{1}{s^2+1}$

$$= \mathrm{e}^{-2s}.\frac{1}{s^2+1} = \frac{\mathrm{e}^{-2s}}{s^2+1}$$

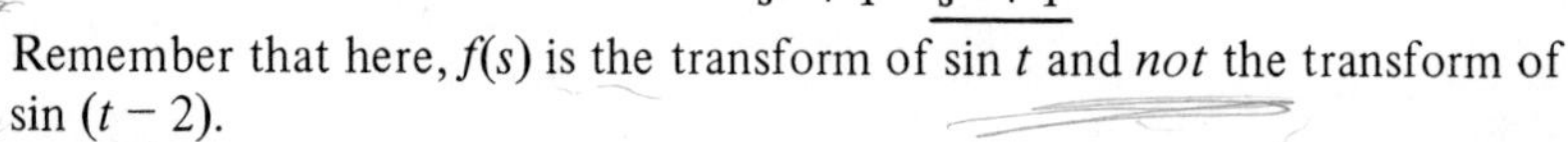
Remember that here, $f(s)$ is the transform of $\sin t$ and *not* the transform of $\sin(t-2)$.

In the same way:

$\mathcal{L}\{\mathrm{H}(t-4).(t-4)^2\} = \mathrm{e}^{-4s}.f(s) \quad \text{where} \quad f(s)$ is the transform of

....

18

$$\boxed{t^2}$$

$$\mathcal{L}\{H(t-4).(t-4)^2\} = e^{-4s}.\mathcal{L}\{t^2\}$$

$$= e^{-4s}\left\{\frac{2!}{s^3}\right\} = \underline{\frac{2\,e^{-4s}}{s^3}}$$

So now you can do these:

(1) $\mathcal{L}\{H(t-5).e^{(t-5)}\} = \ldots\ldots\ldots$

(2) $\mathcal{L}\{H(t-2).\cos(t-2)\} = \ldots\ldots\ldots$

(3) $\mathcal{L}\{H(t-1).(t-1)^4\} = \ldots\ldots\ldots$

(4) $\mathcal{L}\{H(t-\pi/3).\sin 2(t-\pi/3)\} = \ldots\ldots\ldots$

When you have finished all four, move on to frame 19.

19

Results:

(1) $\mathcal{L}\{H(t-5).e^{(t-5)}\} = e^{-5s}.f(s) \qquad f(s) = \mathcal{L}\{e^t\} = \dfrac{1}{s-1}$

$$= \underline{\frac{e^{-5s}}{s-1}}$$

(2) $\mathcal{L}\{H(t-2).\cos(t-2)\} = e^{-2s}.f(s) \qquad f(s) = \mathcal{L}\{\cos t\} = \dfrac{s}{s^2+1}$

$$= \underline{\frac{s\,e^{-2s}}{s^2+1}}$$

(3) $\mathcal{L}\{H(t-1).(t-1)^4\} = e^{-s}.f(s) \qquad f(s) = \mathcal{L}\{t^4\} = \dfrac{4!}{s^5}$

$$= \underline{\frac{24\,e^{-s}}{s^5}}$$

(4) $\mathcal{L}\{H(t-\pi/3).\sin 2(t-\pi/3)\} = e^{-\pi s/3}.f(s) \qquad f(s) = \mathcal{L}\{\sin 2t\} = \dfrac{2}{s^2+4}$

$$= \underline{\frac{2\,e^{-\pi s/3}}{s^2+4}}$$

All correct?

On to frame 20.

20

The last important general result we established was:

$$\mathcal{L}\{\mathrm{H}(t-c).\mathrm{F}(t-c)\} = \mathrm{e}^{-cs}.f(s) \quad \text{where} \quad f(s) = \mathcal{L}\{\mathrm{F}(t)\}.$$

We often need to use this in reverse, so here it is –

$$\underline{\text{If } f(s) = \mathcal{L}\{\mathrm{F}(t)\}, \quad \text{then} \quad \mathrm{e}^{-cs}.f(s) = \mathcal{L}\{\mathrm{H}(t-c).\mathrm{F}(t-c)\}}$$

where c is real and positive.

This is known as the *SECOND SHIFT THEOREM.*

Make a note of it in this form. Write it out and then we shall see how we can apply it in finding inverse transforms.

21

$$\boxed{\text{If } f(s) = \mathcal{L}\{\mathrm{F}(t)\}, \quad \text{then} \quad \mathrm{e}^{-cs}.f(s) = \mathcal{L}\{\mathrm{H}(t-c).\mathrm{F}(t-c)\}}$$

Example 1. Find the function whose transform is $\dfrac{\mathrm{e}^{-2s}}{s-5}$.

The numerator is clearly e^{-cs} where $c = 2$ and therefore indicates $\mathrm{H}(t-2)$.

$$\text{Then } \frac{1}{s-5} = f(s) = \mathcal{L}\{\mathrm{e}^{5t}\} \qquad \therefore\ \mathrm{F}(t) = \mathrm{e}^{5t}$$

$$\therefore\ \mathcal{L}^{-1}\left\{\frac{\mathrm{e}^{-2s}}{s-5}\right\} = \underline{\mathrm{H}(t-2).\mathrm{e}^{5(t-2)}}$$

Remember that, in the result, $\mathrm{F}(t)$ is replaced by ....

22

$$\boxed{\mathrm{F}(t-c)}$$

Example 2. Find $\mathcal{L}^{-1}\left\{\dfrac{3\,\mathrm{e}^{-s}}{s^2+9}\right\}$

The numerator contains e^{-s} and $\therefore$ indicates $\mathrm{H}(t-1)$.

$$\frac{3}{s^2+9} = f(s) = \mathcal{L}\{\sin 3t\} \qquad \therefore\ \mathrm{F}(t) = \sin 3t$$

$$\text{so } \therefore\ \mathcal{L}^{-1}\left\{\frac{3\,\mathrm{e}^{-s}}{s^2+9}\right\} = \text{..........}$$

23

$$\mathcal{L}^{-1}\ \frac{3\,e^{-s}}{s^2+9} = H(t-1).\sin 3(t-1)$$

The sin $3(t-1)$ appears in the result, since $F(t)$, i.e. sin $3t$, is replaced by $F(t-1)$, i.e. t is replaced by $(t-1)$ giving sin $3(t-1)$.

Now another:

Example 3. Find $\mathcal{L}^{-1}\left\{\dfrac{s\,e^{-3s}}{s^2+16}\right\}$

The numerator contains e^{-3s} which indicates $H(t-3)$.

Also $\dfrac{s}{s^2+16} = f(s) = \mathcal{L}\{\ldots\ldots\ldots\}$

so that $\mathcal{L}^{-1}\left\{\dfrac{s\,e^{-3s}}{s^2+16}\right\} = \ldots\ldots\ldots$

Finish it off.

24

$$f(s) = \mathcal{L}\{\cos 4t\}$$

$$\therefore\ \mathcal{L}^{-1}\left\{\frac{s\,e^{-3s}}{s^2+16}\right\} = H(t-3).\cos 4(t-3)$$

Now do these on your own:

(i) $\mathcal{L}^{-1}\left\{\dfrac{e^{-4s}}{s^2}\right\} = \ldots\ldots\ldots$

(ii) $\mathcal{L}^{-1}\left\{\dfrac{3\,e^{-5s}}{s}\right\} = \ldots\ldots\ldots$

(iii) $\mathcal{L}^{-1}\left\{\dfrac{2\,e^{-3s}}{s^3}\right\} = \ldots\ldots\ldots$

When you have completed them, check your results with the next frame.

25

Here are the results in detail.

(i) $\mathcal{L}^{-1}\left\{\dfrac{e^{-4s}}{s^2}\right\} = ?$ $\qquad e^{-4s}$ indicates $H(t-4)$

Also $\mathcal{L}^{-1}\left\{\dfrac{1}{s^2}\right\} = t \qquad \therefore\ F(t) = t$

$\mathcal{L}^{-1}\left\{\dfrac{e^{-4s}}{s^3}\right\} = H(t-4).(t-4).$

(ii) $\mathcal{L}^{-1}\left\{\dfrac{3\,e^{-5s}}{s}\right\} = ?$ $\qquad e^{-5s}$ indicates $H(t-5)$

Also $\mathcal{L}^{-1}\left\{\dfrac{3}{s}\right\} = 3\,\mathcal{L}^{-1}\left\{\dfrac{1}{s}\right\} = 3 \qquad \therefore\ F(t) = 3$

$\therefore\ \mathcal{L}^{-1}\left\{\dfrac{3\,e^{-5s}}{s}\right\} = H(t-5).3 = 3.H(t-5).$

(iii) $\mathcal{L}^{-1}\left\{\dfrac{2\,e^{-3s}}{s^3}\right\} = ?$ $\qquad e^{-3s}$ indicates $H(t-3)$

$\mathcal{L}^{-1}\left\{\dfrac{2}{s^3}\right\} = t^2 \qquad \therefore\ F(t) = t^2$

$\therefore\ \mathcal{L}^{-1}\left\{\dfrac{2\,e^{-3s}}{s^3}\right\} = H(t-3).(t-3)^2.$

Next frame.

26

We can now deal with some rather more interesting examples, but before we do so, let us just list our important results together:

1. $\left.\begin{aligned} H(t-c) &= 0 \quad \text{for} \quad 0 < t < c \\ &= 1 \quad \text{for} \quad t \geqslant c. \end{aligned}\right]$ ---------------- I

2. $\left.\begin{aligned} \mathcal{L}\{H(t-c)\} &= \frac{e^{-cs}}{s} \\ \mathcal{L}\{H(t)\} &= \frac{1}{s}. \end{aligned}\right]$ ---------------------- II

3. $\mathcal{L}\{H(t-c)\,.\,F(t-c)\} = e^{-cs}\,.\,f(s)$ where $f(s) = \mathcal{L}\{F(t)\}.\,]$ ---- III

4. If $f(s) = \mathcal{L}\{F(t)\}$, then $e^{-cs}.f(s) = \mathcal{L}\{H(t-c).F(t-c)\}.]$ ---- IV

These are really all we need for finding Laplace transforms and inverse transforms of functions involving the unit step function.

So on now to frame 27.

27

Example 1. If $F(t) = \mathcal{L}^{-1}\left\{\frac{2}{s} - \frac{3\,e^{-s}}{s^2} + \frac{3\,e^{-4s}}{s^2}\right\}$

find $F(t)$ and sketch the graph of the function.

We take each term in turn and find its inverse transform.

1. $\mathcal{L}^{-1}\left\{\frac{2}{s}\right\} = 2\,\mathcal{L}^{-1}\left\{\frac{1}{s}\right\} = 2.$

(ii) $\mathcal{L}^{-1}\left\{\frac{3\,e^{-s}}{s^2}\right\} = H(t-1).3(t-1)$

(iii) $\mathcal{L}^{-1}\left\{\frac{3\,e^{-4s}}{s^2}\right\} =$ ....

28

$$\mathcal{L}^{-1}\left\{\frac{3\,e^{-4s}}{s^2}\right\} = H(t-4).3(t-4)$$

So we have:

$$\mathcal{L}^{-1}\left\{\frac{2}{s}\right\} = 2; \quad \mathcal{L}^{-1}\left\{\frac{3\,e^{-s}}{s^2}\right\} = H(t-1).3(t-1);$$

and

$$\mathcal{L}^{-1}\left\{\frac{3\,e^{-4s}}{s^2}\right\} = H(t-4).3(t-4).$$

$$F(t) = \mathcal{L}^{-1}\left\{\frac{2}{s} - \frac{3\,e^{-s}}{s^2} + \frac{3\,e^{-4s}}{s^2}\right\}$$

$$\therefore\ \underline{F(t) = 2H(t) - H(t-1).3(t-1) + H(t-4).3(t-4)}$$

Remembering the definition of $H(t-c)$, what is the value of $F(t)$ between $t = 0$ and $t = 1$?

29

$$F(t) = 2$$

Indeed it is, since between $t = 0$ and $t = 1$,

$$H(t) = 1, \quad \text{but} \quad H(t-1) = 0 \quad \text{and} \quad H(t-4) = 0.$$

Similarly between $t = 1$ and $t = 4$

$F(t) = \dots\dots\dots\dots\dots\dots\dots\dots$

30

$$F(t) = -3t + 5$$

Between $t = 1$ and $t = 4$, $H(t) = 1$, $H(t-1) = 1$, but $H(t-4) = 0$.

$$\therefore\ F(t) = 2 - 3(t-1)$$

$$= 2 - 3t + 3 \qquad \therefore\ F(t) = -3t + 5.$$

And, in like fashion, for $t > 4$,

$F(t) = \dots\dots\dots\dots\dots\dots\dots\dots$

31

$$\boxed{F(t) = -7}$$

For $t > 4$, $H(t) = 1$; $\quad H(t-1) = 1$; $\quad H(t-4) = 1$.

$$\therefore\ F(t) = 2 - 3(t-1) + 3(t-4)$$

$$= 2 - 3t + 3 + 3t - 12 = -7 \qquad \therefore\ F(t) = -7$$

So, collecting results, we have:

For			
	$0 < t < 1$	$F(t) = 2$	
	$1 < t < 4$	$F(t) = -3t + 5$	$\begin{cases} t = 1, F(t) = 2 \\ t = 4, F(t) = -7 \end{cases}$
	$4 < t$	$F(t) = -7$.	

Now use these facts to sketch the graph of $F(t)$ between $t = 0$ and $t = 5$.

32

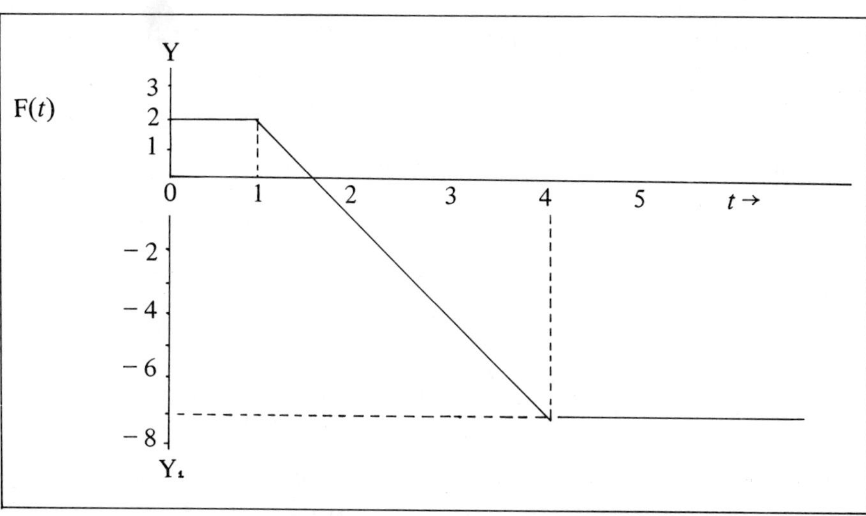

There it is. Now do this one:

Example 2. If $F(t) = \mathcal{L}^{-1}\left\{\dfrac{3}{s} + \dfrac{2\,e^{-2s}}{s^2} - \dfrac{2\,e^{-5s}}{s^2}\right\}$, find $F(t)$

and sketch the graph of the function.

Deal with the terms one at a time. So first we have

$$\mathcal{L}^{-1}\left\{\frac{3}{s}\right\} = \ldots\ldots\ldots\ldots\ldots\ldots\ldots\ldots\ldots\ldots$$

33

$$\mathcal{L}^{-1}\left\{\frac{3}{s}\right\} = 3\,\mathrm{H}(t)$$

Now the second one.

$$\mathcal{L}^{-1}\left\{\frac{2\,\mathrm{e}^{-2s}}{s^2}\right\} = \ldots\ldots\ldots\ldots$$

34

$$\mathcal{L}^{-1}\left\{\frac{2\,\mathrm{e}^{-2s}}{s^2}\right\} = \mathrm{H}(t-2)\,.\,2(t-2)$$

since the e^{-2s} indicates $\mathrm{H}(t-2)$.

$$\mathcal{L}^{-1}\left\{\frac{2}{s^2}\right\} = 2t \qquad \therefore\ \mathrm{F}(t) = 2t \qquad \therefore\ \mathrm{F}(t-2) = 2(t-2)$$

$$\therefore\ \mathcal{L}^{-1}\left\{\frac{2\,\mathrm{e}^{-2s}}{s^2}\right\} = \mathrm{H}(t-2)\,.\,2(t-2).$$

Now there just remains the third term to deal with.

$$\mathcal{L}^{-1}\left\{\frac{2\,\mathrm{e}^{-5s}}{s^2}\right\} = \ldots\ldots\ldots\ldots$$

35

$$\mathcal{L}^{-1}\left\{\frac{2\,\mathrm{e}^{-5s}}{s^2}\right\} = \mathrm{H}(t-5).2(t-5)$$

$\therefore$ Collecting the three results together, with appropriate signs:

$$\underline{\mathrm{F}(t) = 3.\mathrm{H}(t) + \mathrm{H}(t-2).2(t-2) - \mathrm{H}(t-5).2(t-5).}$$

We notice that the function starts at $t = 0$ and that there are 'break points' at $t = 2$ and $t = 5$.

We investigate what happens between these values of t.

For $0 < t < 2$, $\quad \mathrm{F}(t) = \ldots\ldots\ldots\ldots$

$2 < t < 5$, $\quad \mathrm{F}(t) = \ldots\ldots\ldots\ldots$

$5 < t$ $\quad \mathrm{F}(t) = \ldots\ldots\ldots\ldots$

36

$0 < t < 2,$	$F(t) = 3$
$2 < t < 5,$	$F(t) = 2t - 1$
$5 < t,$	$F(t) = 9$

Here is the working, just as a check:

$0 < t < 2, \quad F(t) = 3.1 + 0 - 0 \qquad \therefore \underline{F(t) = 3}$

$2 < t < 5, \quad F(t) = 3.1 + 1.2(t - 2) - 0 \qquad \therefore \underline{F(t) = 2t - 1}$

$5 < t, \quad F(t) = 3.1 + 1.2(t - 2) - 1.2(t - 5) \qquad \therefore \underline{F(t) = 9}$

Next frame.

37

Before we sketch the graph, check the function values at the 'break points'

(i) $F(t) = 3 \qquad \therefore$ At $t = 2$, $\quad F(t) = 3$

(ii) $F(t) = 2t - 1 \qquad \therefore$ At $t = 2$, $\quad F(t) = 2.2 - 1 = 3$
At $t = 5$, $\quad F(t) = 2.5 - 1 = 9$

(iii) $F(t) = 9 \qquad \therefore$ At $t = 5$, $\quad F(t) = 9$

This step is important, since there may be discontinuities where the function suddenly changes.

Now with this information, you can sketch the graph over this range.

Do that.

38

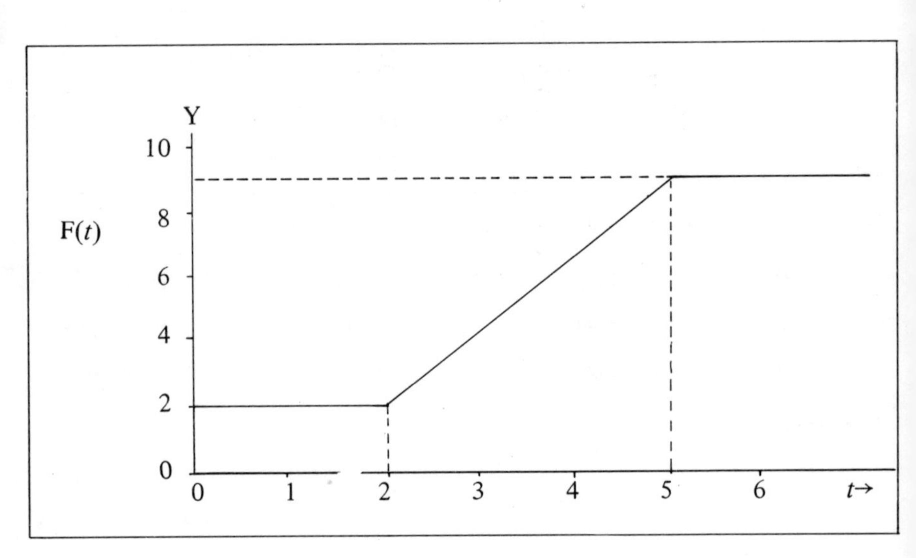

In this example, we see that the three separate functions link up at the 'break points' to give a continuous line. This is not always the case as we shall see later.

Move on now to the next frame for another example.

39

Example 3. $\mathrm{F}(t) = \mathcal{L}^{-1}\left\{\dfrac{(1-\mathrm{e}^{-s})(1+\mathrm{e}^{-2s})}{s^2}\right\}$

Find $\mathrm{F}(t)$ and sketch the graph of the function between $t = 0$ and $t = 5$.

First multiply out the numerator into separate terms.

$$\mathrm{F}(t) = \mathcal{L}^{-1}\left\{\frac{1-\mathrm{e}^{-s}+\mathrm{e}^{-2s}-\mathrm{e}^{-3s}}{s^2}\right\}$$

$$= \mathcal{L}^{-1}\left\{\frac{1}{s^2} - \frac{\mathrm{e}^{-s}}{s^2} + \frac{\mathrm{e}^{-2s}}{s^2} - \frac{\mathrm{e}^{-3s}}{s^2}\right\}$$

Now write down the inverse transform of each term, so that

$\mathrm{F}(t) = \ldots\ldots\ldots\ldots\ldots\ldots\ldots\ldots$

40

$$F(t) = t.H(t) - H(t-1).(t-1) + H(t-2).(t-2) - H(t-3).(t-3)$$

$\therefore$ For

$0 < t < 1$,	$F(t) = t - 0 + 0 - 0$	$\therefore$ $F(t) = t$
$1 < t < 2$,	$F(t) = t - (t-1) + 0 - 0$	$\therefore$ $F(t) = 1$
$2 < t < 3$,	$F(t) = t - (t-1) + (t-2) - 0$	$\therefore$ $F(t) = t - 1$
$3 < t$,	$F(t) = t - (t-1) + (t-2) - (t-3)$	$\therefore$ $F(t) = 2$

The second and fourth of these component functions are constants, but before sketching the graph we must check the values of $F(t) = t$ and $F(t) = t - 1$ at the appropriate break points.

What are these values?

41

(i)	$F(t) = t$	At $t = 1$,	$F(t) = 1$
(ii)	$F(t) = t - 1$.	At $t = 2$,	$F(t) = 1$
		$t = 3$,	$F(t) = 2$

Now we can go ahead and sketch the graph. What does it look like?

42

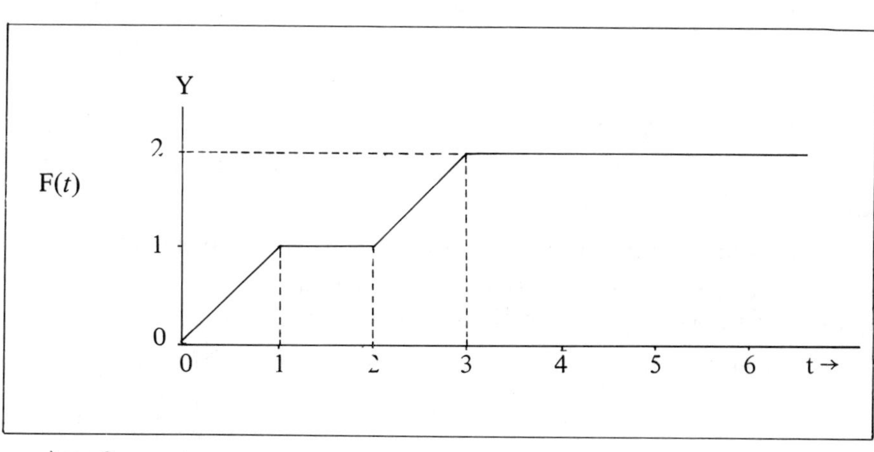

Agree?

Now another example, so move on to frame 43.

43

Example 4. A function $F(t)$ is defined by

$$F(t) = 3 \quad \text{for} \quad 0 < t < 1$$
$$= t \quad \text{for} \quad 1 < t$$

Sketch the graph of the function and find its Laplace transform.

In this case, we are working in the reverse order. We are given the function and have to find its Laplace transform. First we sketch the graph.

Notice that

(i) Between $t = 0$ and $t = 1$, $F(t) = 3$, i.e. constant.

(ii) From $t = 1$ onwards, $F(t) = t$, i.e. the graph is a straight line with unit slope. At the break point $t = 1$, $F(t) = 1$.

Now sketch the graph.

44

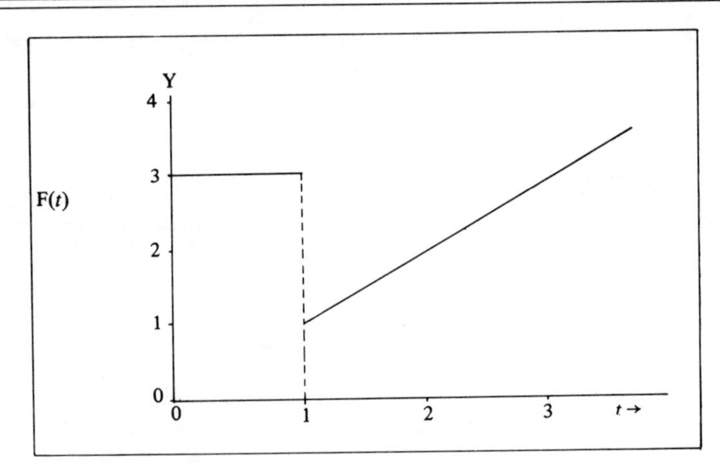

Note the discontinuity at $t = 1$.

So much for the graph. Now we have to express $F(t)$ in function terms.

(i) Between $t = 0$ and $t = 1$, $F(t)$ is constant and equals 3.
(ii) At $t = 1$, we must cancel $F(t) = 3$ and replace it by $F(t) = t$

$$\therefore \; F(t) = 3.H(t) \quad - \quad 3.H(t-1) \quad + \quad H(t-1).t$$

$3.H(t)$	$3.H(t-1)$	$H(t-1).t$
from $t = 0$ onwards	$F(t) = 3$ switched off at $t = 1$	$F(t) = t$ switched on at $t = 1$

Be quite sure that you understand that step. It is vitally important. When you are quite happy about it, continue on to the next frame.

45

So we have:

$$F(t) = 3.H(t) - 3.H(t-1) + H(t-1).t$$

and we have to express this in Laplace transforms. We can easily deal with the first two terms, but the third term is not yet in the right form to enable us to write its transform.

For remember $\mathcal{L}\{H(t-c).F(t-c)\} = e^{-cs}.f(s)$ where $f(s) = \mathcal{L}\{F(t)\}$

$\therefore$ We must re-write the third term in this way:

$$H(t-1).t = H(t-1).(t-1) + 1.H(t-1).$$

We now have

$$F(t) = 3.H(t) - 3.H(t-1) + H(t-1).(t-1) + 1.H(t-1)$$

i.e. $$F(t) = 3.H(t) - 2.H(t-1) + H(t-1).(t-1)$$

$\therefore \mathcal{L}\{F(t)\} = \ldots\ldots\ldots\ldots$

Complete the job, taking the three terms one at a time.

46

$$\mathcal{L}\{F(t)\} = \frac{3}{s} - \frac{2e^{-s}}{s} + \frac{e^{-s}}{s^2}$$

Example 5. For the function $\left\{\begin{aligned} F(t) &= 5 & 0<t<3 \\ &= 2t+1, & 3<t \end{aligned}\right\}$, sketch the graph of the function and find its Laplace transform.

Note that (i) For $0<t<3$, $F(t) = 5$ (constant)

(ii) For $3<t$, $F(t) = 2t+1$, and at $t = 3$, $F(t) = 7$.

You can now sketch the graph.

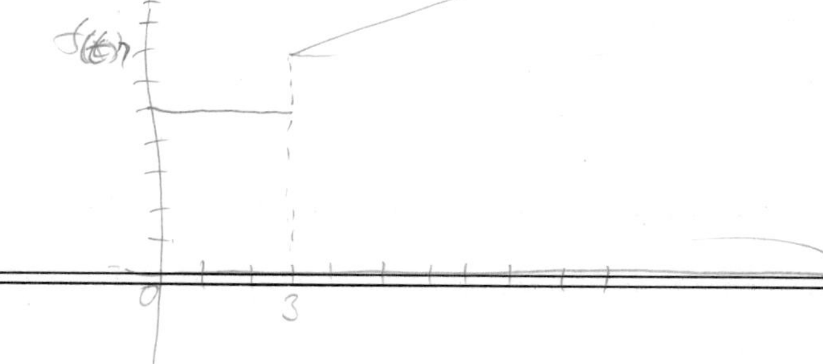

47

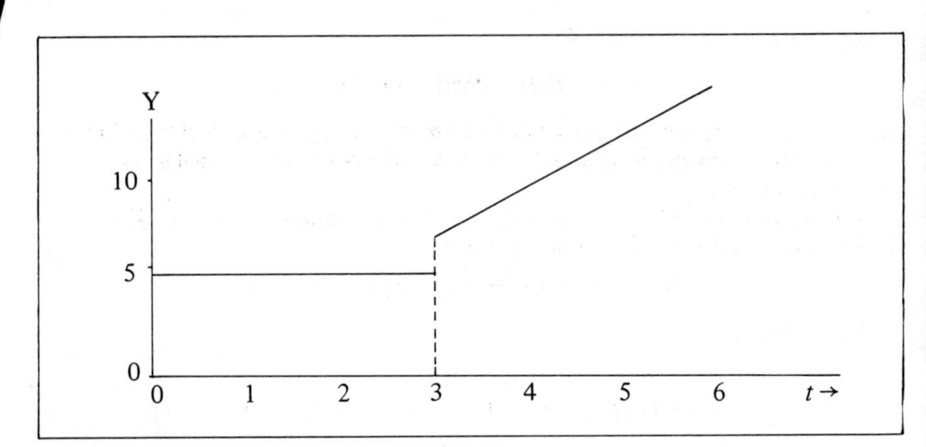

Notice the discontinuity at $t = 3$.

Expressing the function in unit step form

$$F(t) = 5.H(t) - 5.H(t-3) + H(t-3).(2t+1)$$

The third term must be re-written as a function of $(t-3)$ to agree with the $H(t-3)$, with an extra term added as necessary to adjust the expression.

i e. $$H(t-3).(2t+1) = H(t-3).2(t-3) + 7.H(t-3)$$

$$\therefore\ F(t) = 5.H(t) - 5.H(t-3) + H(t-3).2(t-3) + 7.H(t-3)$$

$$\therefore\ F(t) = 5.H(t) + 2.H(t-3) + H(t-3).2(t-3)$$

$$\therefore\ \mathcal{L}\{F(t)\} = \ldots\ldots\ldots\ldots\ldots\ldots$$

48

$$\mathcal{L}\{F(t)\} = \frac{5}{s} + \frac{2\,e^{-3s}}{s} + \frac{2\,e^{-3s}}{s^2}$$

Now what about this one?

Example 6. If $\mathcal{L}\{F(t)\} = \dfrac{(1 + e^{-2s})^2}{s^2}$

express $F(t)$ in terms of the unit step function and hence evaluate $F(1)$, $F(3)$, $F(5)$.

First we have to find $F(t)$.

$$\mathcal{L}\{F(t)\} = \frac{(1 + e^{-2s})^2}{s^2} = \frac{1 + 2\,e^{-2s} + e^{-4s}}{s^2}$$

$$= \frac{1}{s^2} + \frac{2\,e^{-2s}}{s^2} + \frac{e^{-4s}}{s^2}$$

$\therefore$ $F(t) =$ ....

49

$$F(t) = t.H(t) + 2.H(t-2).(t-2) + H(t-4).(t-4)$$

Now we have to evaluate $F(1)$, $F(3)$, $F(5)$.

For $0 < t < 2$,	$F(t) = t + 0 + 0$	$F(1) = 1$
$2 < t < 4$,	$F(t) =$	$F(3) =$ ....
$4 < t$,	$F(t) =$	$F(5) =$ ....

Check with the results in frame 50.

50

$2 < t < 4,$	$F(t) = t + 2(t-2) + 0$ $= 3t - 4$	$\therefore$ $\underline{F(3) = 5}$
$4 < t,$	$F(t) = t + 2(t-2) + (t-4)$ $= 4t - 8$	$\therefore$ $\underline{F(5) = 12}$

So the results are: $\underline{F(1) = 1; \quad F(3) = 5; \quad F(5) = 12.}$

Now one slightly more involved.

Example 7. A function F(t) is defined by

$$\begin{aligned} F(t) &= t^2 && 0 < t < 2 \\ &= 4 && 2 < t < 5 \\ &= 0 && 5 < t \end{aligned}$$

Sketch the graph of the function and find its Laplace transform.

First of all, sketch the graph. Off you go on your own.

51

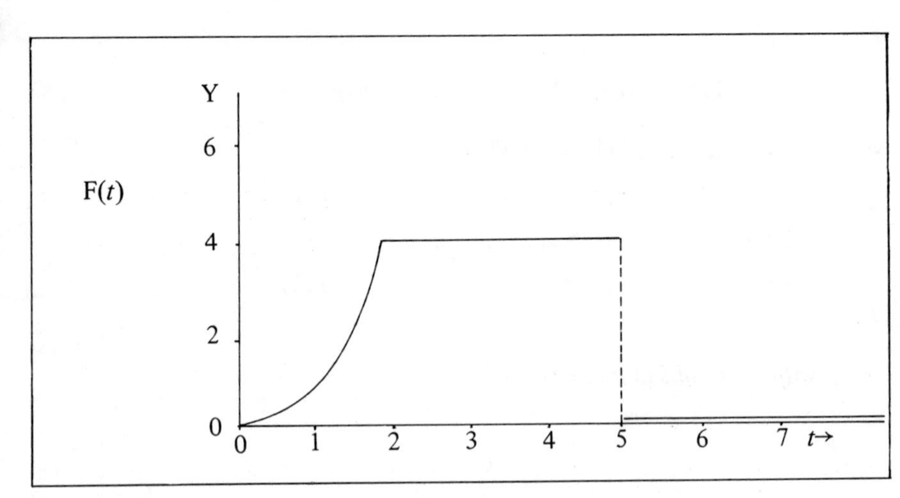

Now interpret the graph in terms of the unit step function.

$F(t) =$ .... $t^2H(t) - t^2H(t-2) + 4H(t-2) - 4H(t-5)$

52

$$\boxed{F(t) = t^2 . H(t) - t^2 . H(t-2) + 4 . H(t-2) - 4 . H(t-5)}$$

This indicates clearly the 'switching-on' and 'switching-off' of the various terms. Before we can take Laplace transforms, however, terms of the form $H(t-c).F(t)$ must be expressed in the form $H(t-c).F(t-c)$ with any extra terms added or subtracted to adjust the expression.

e.g. $\quad t^2 . H(t-2) = (t-2)^2 . H(t-2) + \ldots\ldots\ldots\ldots$

53

$$t^2 . H(t-2) = (t-2)^2 . H(t-2) + \boxed{(4t-4) . H(t-2)}$$

since $(t-2)^2 . H(t-2) = (t^2 - 4t + 4) . H(t-2)$

$$\therefore \; t^2 . H(t-2) = (t-2)^2 . H(t-2) + (4t-4) . H(t-2).$$

But the second term of *this* expression must also be adjusted in the same way.

$$(4t-4) . H(t-2) = 4(t-2) . H(t-2) + 4 . H(t-2)$$

$$\therefore \; t^2 . H(t-2) = (t-2)^2 . H(t-2) + 4(t-2) . H(t-2) + 4 . H(t-2)$$

If we now use this expression to amend the original function, we get

$F(t) = \ldots\ldots\ldots\ldots\ldots\ldots\ldots\ldots$

54

We get:

$$F(t) = t^2 . H(t) - (t-2)^2 . H(t-2) + 4(t-2) . H(t-2)$$
$$+ 4 . H(t-2) + 4 . H(t-2) - 4 . H(t-5)$$

which tidies up to become

$$\boxed{F(t) = t^2 . H(t) - (t-2)^2 . H(t-2) + 4(t-2) . H(t-2) + 8 . H(t-2) - 4 . H(t-5)}$$

Now at last we can write the Laplace transform of the entire function.

$\mathcal{L}\{F(t)\} = \ldots\ldots\ldots\ldots\ldots\ldots\ldots\ldots$

55

$$\mathcal{L}\{F(t)\} = \frac{2}{s^3} - \frac{2\,e^{-2s}}{s^3} + \frac{4\,e^{-2s}}{s^2} + \frac{8\,e^{-2s}}{s} - \frac{4\,e^{-5s}}{s}$$

Agree?

Remember that terms of the form $H(t-c).F(t)$ must be converted into the form $H(t-c).F(t-c)$ before the transform can be written.

Now here is one that requires slightly different treatment.

Example 8. Find the Laplace transform of the function defined by

$$\begin{aligned} F(t) &= e^{2t} \qquad 0<t<2 \\ &= 0 \qquad 2<t \end{aligned}$$

First we will sketch the graph of the function. You do that.

56

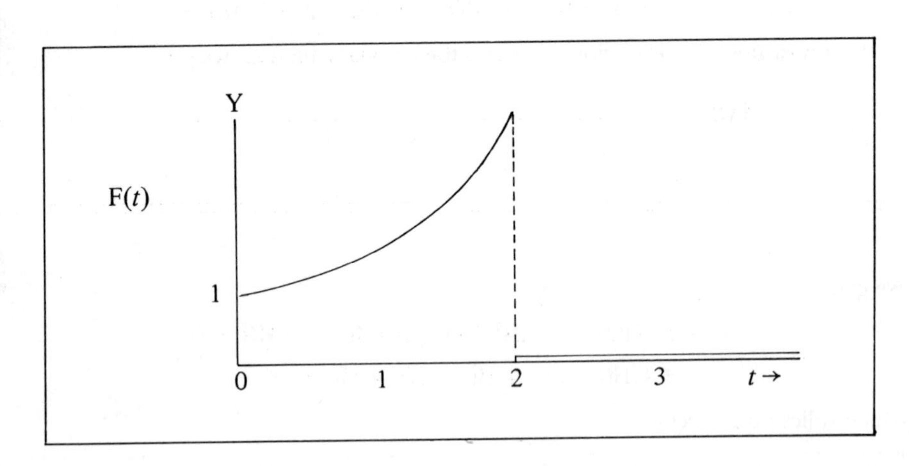

Then, in terms of the step function –

F(t) = ...

57

$$\boxed{F(t) = e^{2t}.H(t) - e^{2t}.H(t-2)}$$

Next we must rewrite the second term so that e^{2t} is expressed in terms of $(t - 2)$ in order to match up with $H(t - 2)$.

$$\therefore\ e^{2t}.H(t-2) = e^{2(t-2)}.H(t-2) + \ldots\ldots ? \ldots\ldots$$

The question now is how must we adjust $e^{2(t-2)}$ to bring it back to e^{2t}? Clearly adding or subtracting terms will not do the trick. What then?

When you have thought about it, turn on to see one way of tackling it.

58

$$e^{2(t-2)} = e^{2t-4} = e^{2t}.e^{-4}$$

$\therefore$ We must divide by e^{-4} (or multiply by e^4) to bring $e^{2(t-4)}$ back to e^{2t}

So we have this:

$$\begin{aligned} F(t) &= e^{2t}.H(t) - e^{2t}.H(t-2) \\ &= e^{2t}.H(t) - e^{2(t-2)}.H(t-2).e^4 \\ &= e^{2t}.H(t) - e^4.e^{2(t-2)}.H(t-2) \end{aligned}$$

Since e^4 is simply a constant factor in the expression, we can now write the transforms.

$$\mathcal{L}\{F(t)\} = \ldots\ldots\ldots\ldots\ldots\ldots$$

59

$$\mathcal{L}\{F(t)\} = \frac{1}{s-2} - e^4 . \frac{1}{s-2} . e^{-2s}$$

$$\boxed{\mathcal{L}\{F(t)\} = \frac{1}{s-2}\,[1 - e^{4-2s}]}$$

Note: We could, of course, have found the Laplace transform from first principles – and sometimes we have to do so.

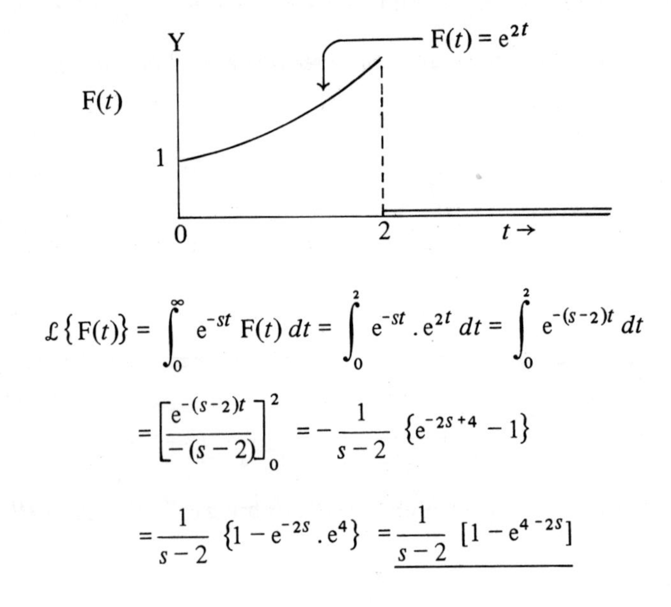

$$\mathcal{L}\{F(t)\} = \int_0^\infty e^{-st}\,F(t)\,dt = \int_0^2 e^{-st} . e^{2t}\,dt = \int_0^2 e^{-(s-2)t}\,dt$$

$$= \left[\frac{e^{-(s-2)t}}{-(s-2)}\right]_0^2 = -\frac{1}{s-2}\,\{e^{-2s+4} - 1\}$$

$$= \frac{1}{s-2}\,\{1 - e^{-2s} . e^4\} = \underline{\frac{1}{s-2}\,[1 - e^{4-2s}]}$$

Now on to the next.

60

Example 9. Find the Laplace transform of the function F(*t*), such that

$$F(t) = \sin 2t \qquad 0 < t < \pi$$
$$= 0 \qquad \pi < t$$

As always, we first sketch the graph of the function.

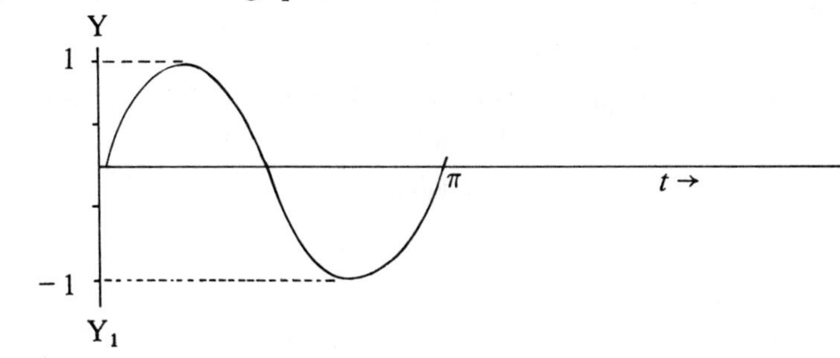

If we do this one from first principles, we get

$$\mathcal{L}\{F(t)\} = \int_0^\infty e^{-st}\, F(t)\, dt = \int_0^\pi e^{-st} . \sin 2t\, dt, \qquad \text{since } F(t) = 0 \text{ for } t > \pi.$$

Now $\qquad e^{j\theta} = \cos\theta + j\sin\theta \qquad \sin\theta = \mathcal{I}\{e^{j\theta}\} \qquad \therefore \sin 2t = \mathcal{I}\{e^{j2t}\}$

$$\therefore \mathcal{L}\{F(t)\} = \mathcal{I}\int_0^\pi e^{-st}\, e^{j2t}\, dt = \mathcal{I}\int_0^\pi e^{-(s-j2)t}\, dt$$

$$= \mathcal{I}\left[\frac{e^{-(s-j2)t}}{-(s-j2)}\right]_0^\pi = -\mathcal{I}\frac{1}{s-j2}\left\{e^{-(s-j2)\pi} - 1\right\}$$

$$= \mathcal{I}\frac{1}{s-j2}\left\{1 - e^{-s\pi}\, e^{j2\pi}\right\}$$

$$= \mathcal{I}\frac{s+j2}{s^2+4}\left\{1 - e^{-s\pi}\right\} \qquad \text{since } e^{j2\pi} = \cos 2\pi + j\sin 2\pi = 1 + j0 = 1$$

$$\therefore \mathcal{L}\{F(t)\} = \frac{2}{s^2+4}\left\{1 - e^{-s\pi}\right\}$$

OR We could have used the unit step function in this example.

$$F(t) = \sin 2t . H(t) - \sin 2t . H(t-\pi)$$

which must then be written $F(t) = \sin 2t . H(t) - \sin 2(t-\pi) . H(t-\pi) + ?$
How can we adjust the expression in this case? It is different: think about it and use your knowledge of trig. identities.

61

We had

$$F(t) = \sin 2t\,.\,H(t) - \sin 2(t-\pi)\,.\,H(t-\pi) + ?$$

Now $\sin 2(t-\pi) = \sin(2t - 2\pi)$

$= \sin 2t.$

In this particular example, it so happens that $\sin 2(t-\pi)$ is equivalent to $\sin 2t$, so that, in fact, no adjustment to the expression is required.

$$\therefore\ F(t) = \sin 2t\,.\,H(t) - \sin 2(t-\pi)\,.\,H(t-\pi)$$

$$\therefore\ \mathcal{L}\{F(t)\} = \dots\dots\dots\dots$$

62

$$\mathcal{L}\{F(t)\} = \frac{2}{s^2+4} - \frac{2\,e^{-\pi s}}{s^2+4} = \frac{2}{s^2+4}\{1 - e^{-\pi s}\}$$

And now, one final example.

Example 10. Find the Laplace transform of the function whose graph is shown.

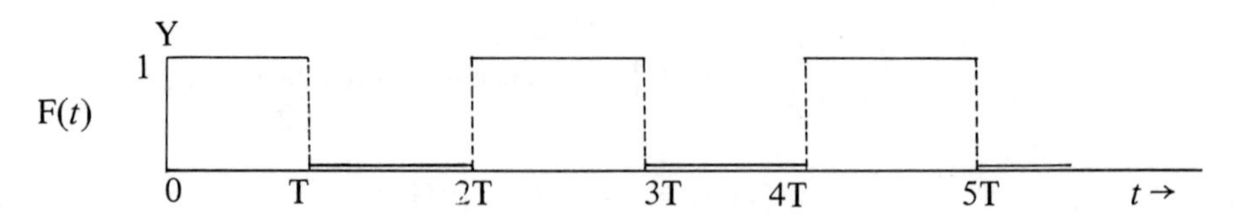

This is a periodic function.

$$F(t) = 1 \quad 0 < t < T$$
$$= 0 \quad T < t < 2T$$
$$= 1 \quad 2T < t < 3T \quad \text{etc.}$$

$$\therefore\ F(t) = 1\,.\,H(t) - 1\,.\,H(t-T) + 1\,.\,H(t-2T) - 1\,.\,H(t-3T) + \dots$$

$$\therefore\ \mathcal{L}\{F(t)\} = \dots\dots\dots\dots$$

63

$$\mathcal{L}\{F(t)\} = \frac{1}{s} - \frac{e^{-Ts}}{s} + \frac{e^{-2Ts}}{s} - \frac{e^{-3Ts}}{s} + \ldots$$

$$\therefore \mathcal{L}\{F(t)\} = \frac{1}{s}[1 - e^{-Ts} + e^{-2Ts} - e^{-3Ts} + \ldots]$$

You may remember that

$$(1+x)^{-1} = 1 - x + x^2 - x^3 + x^4 - \ldots$$

so that the infinite series within the brackets is the expansion of

....

64

$$\{1 + e^{-Ts}\}^{-1}$$

$$\therefore \mathcal{L}\{F(t)\} = \frac{1}{s}\{1 + e^{-Ts}\}^{-1} = \frac{1}{s(1 + e^{-Ts})}$$

$$\therefore \mathcal{L}\{F(t)\} = \frac{1}{s(1 + e^{-Ts})}$$

Periodic functions form an important class in their own right, so we shall have a special look at them in the next programme.

This, then, is the end of the programme dealing with the Heaviside Unit Step Function and some of its applications.

Before working through the Text Exercise, read down the Summary Sheet set out in the next frame.

65

Revision Summary

1. *Heaviside unit step function:* $H(t-c)$

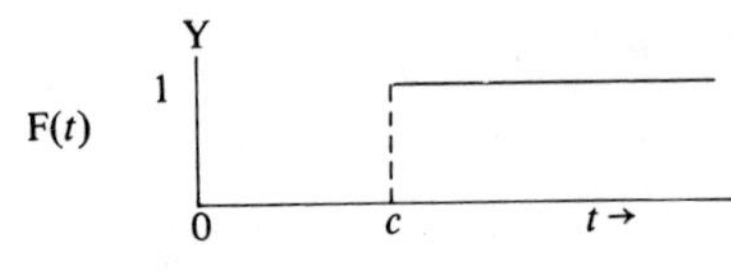

$$F(t) = 0 \quad 0 < t < c$$
$$ = 1 \quad c < t$$

2.

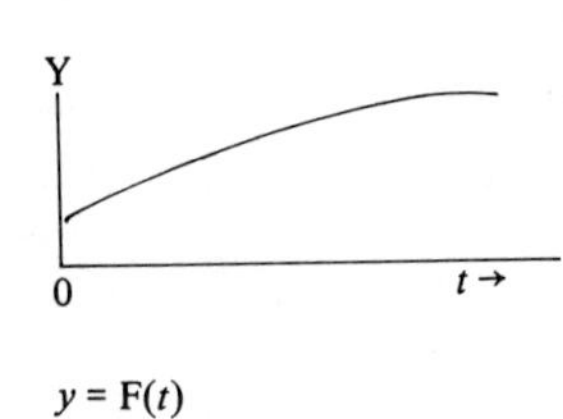

$y = F(t)$

$y = F(t).H(t-c)$

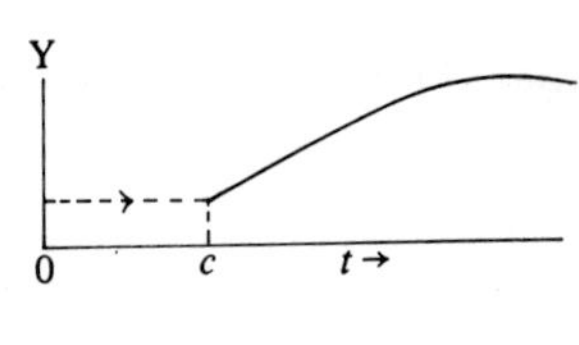

$y = F(t-c).H(t-c)$

3. *Laplace transform of* $H(t-c)$

$$\mathcal{L}\{H(t-c)\} = \frac{e^{-cs}}{s}$$

$$\mathcal{L}\{F(t)\} = \frac{1}{s}$$

4. *Laplace transform of* $H(t-c).F(t-c)$

$$\mathcal{L}\{H(t-c).F(t-c)\} = e^{-cs}.f(s) \qquad \text{where.} \qquad f(s) = \mathcal{L}\{F(t)\}$$

5. *Second shift theorem*

If $f(s) = \mathcal{L}\{F(t)\}$, then $e^{-cs}.f(s) = \mathcal{L}\{H(t-c).F(t-c)\}$ where c is real and positive.

....

If there is any point of the work on which you are not perfectly clear, go back and revise it. There is no hurry. When you are ready, turn on to the Test Exercise.

66

The problems below are all straightforward and based directly on the work of this programme. Do all the questions, but take your time. There is no need to hurry over the exercise.

Test Exercise–V

1. In each of the following cases, sketch the graph of the function and find its Laplace transform.

   (i) $F(t) = 2t \qquad 0 < t < 2$
   $\quad\; = 4 \qquad 2 < t$

   (ii) $F(t) = t^2 \qquad 0 < t < 2$
   $\quad\; = t - 1 \qquad 2 < t < 3$
   $\quad\; = 7 \qquad 3 < t$

   (iii) $F(t) = e^{-3t} \qquad 0 < t < 2$
   $\quad\; = 0 \qquad 2 < t$

   (iv) $F(t) = \sin 3t \qquad 0 < t < \pi$
   $\quad\; = 0 \qquad \pi < t.$

2. Find the function $F(t)$ whose transform is

$$f(s) = \frac{3}{s} - \frac{e^{-s}}{s} + \frac{6\,e^{-3s}}{s}$$

   Sketch the graph of the function between $t = 0$ and $t = 5$.

3. If $F(t) = \mathcal{L}^{-1}\left\{\dfrac{(1 - e^{-2s})(1 + 2\,e^{-3s})}{s^2}\right\}$,

   evaluate F(1), F(4), F(6). Sketch the function.

Further Problems – V

1. Determine the Laplace transforms of

   (a) $\sinh 2t \,.\, H(t-3)$
   (b) $\cos t \,.\, H(t-\pi)$
   (c) $t^2 \,.\, H(t-2)$

2. Find the inverse transforms of

   (a) $\dfrac{s}{(s^2+1)(1-e^{-\pi s})}$

   (b) $\dfrac{e^{-2s}}{s^2-6s+13}$

   (c) $\dfrac{e^{-2s}}{(s+3)^3}$

   (d) $\dfrac{1}{s(1-e^{-\pi s})}$

   (e) $\dfrac{e^{-3s}}{(s-4)^3}$

3. Determine the Laplace transform of $F(t) = \sum\limits_{n=0}^{\infty} H(t-n\pi)$ for $n = 1, 2, 3, \ldots,$ expressing the result in its simplest form.

4. Determine the function $F(t)$ for which

   $$F(t) = \mathcal{L}^{-1}\left\{\frac{4}{s} - \frac{5\,e^{-s}}{s^2} + \frac{5\,e^{-3s}}{s^2}\right\}$$

5. The current i in a circuit at a time t is given by

   $$\ddot{i} - 4\dot{i} - 5i = 30\,H(t-2)$$

   Solve the equation for i, given that at $t = 0$, $i = \dot{i} = 0$.

6. If $F(t) = \cos t$ for $0 < t < \pi$
   $= \sin t$ for $t > \pi$
   find $\mathcal{L}\{F(t)\}$.

7. Find $\mathcal{L}\{F(t)\}$ when $F(t) = \sin 2t$ $\quad 0 < t < \pi$
   $= 0$ $\quad t > \pi$

8. Express in terms of the Heaviside unit step function

   (a) $F(t) = t^2 \quad 0 < t < 2$
   $= 4t \quad t > 2$

   (b) $F(t) = \sin t \quad 0 < t < \pi$
   $= \sin 2t \quad \pi < t < 2\pi$
   $= \sin 3t \quad t > 2\pi$

9. Find $\mathcal{L}\{F(t)\}$ where
$$F(t) = \begin{cases} 0 & 0 < t < 1 \\ t & 1 < t < 2 \\ 0 & t > 2 \end{cases}$$

10. If
$$F(t) = \begin{cases} t^2 & 0 < t < 2 \\ t - 1 & 2 < t < 3 \\ 7 & t > 3 \end{cases}$$
determine $\mathcal{L}\{F(t)\}$.

11. Solve the equation
$$\ddot{y} + y = F(t)$$
where $F(t) = n + 1$ for $\pi n < t < (n + 1)\pi$, with $n = 0, 1, 2, \ldots$ and given that $y = \dot{y} = 0$ when $t = 0$.

12. A constant voltage E for the time interval $t = t_1$ to $t = t_2$ is applied to a circuit containing R and C in series. The capacitor at $t = 0$ has a charge q_0. Show that the current at time t is given by
$$i = -\frac{q_0}{RC}\,e^{-t/RC} + \frac{E}{R}\{e^{-(t-t_1)/RC} \,.\, H(t - t_1) - e^{-(t-t_2)/RC} \,.\, H(t - t_2)\}.$$

13. Solve the equation $\ddot{x} + x = F(t)$ where
$$F(t) = \begin{cases} 3 & 0 < t < 4 \\ 2t - 5 & t > 4 \end{cases}$$
and, at $t = 0, x = 1, \dot{x} = 0$.

Programme 6

PERIODIC FUNCTIONS

1

Introduction

In many practical situations, we are dealing with problems of vibrations or oscillations, either mechanical or electrical. These are periodic by nature and it is important that consideration be given to obtaining the Laplace transforms of such functions.

Towards the end of the previous programme, we worked through one such example, using the Heaviside unit step function and summing the resulting infinite series. Sometimes we have to solve a problem in that way: in many cases, however, an easier method is available and this is what we are now going to develop.

2

Laplace transform of a periodic function

Let $F(t)$ be a periodic function of period w, i.e. all values of $F(t)$ occur regularly at intervals of w units of t.

i.e. $$F(t + nw) = F(t).$$

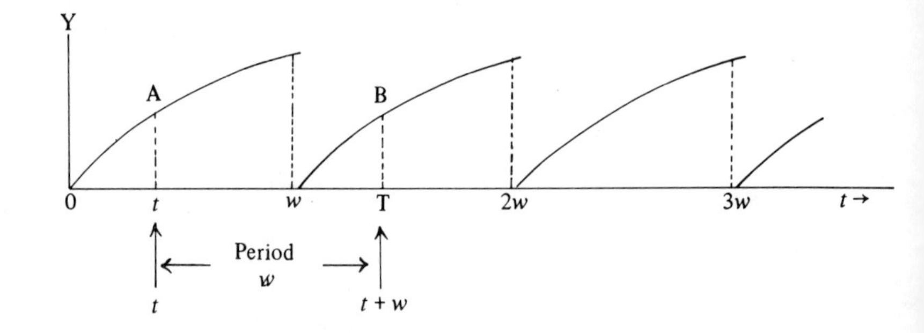

Let A and B be two corresponding points on two successive cycles. Then the values of t at A and B differ by the period.

i.e. $$T = t + w \qquad \therefore\ t = T - w$$

That sets the scene. Now to find $\mathcal{L}\{F(t)\}$.

3

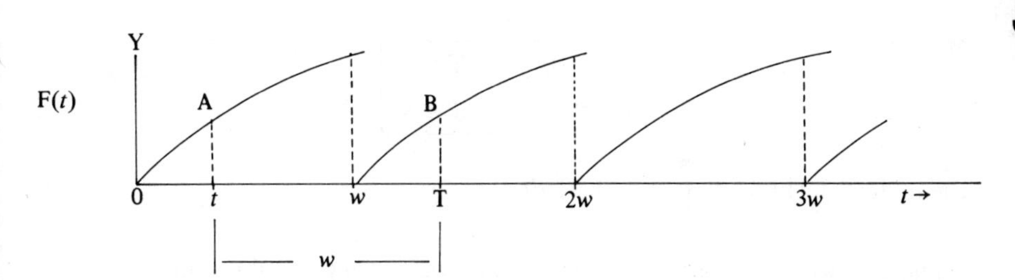

So we have $\quad T = t + w \qquad \therefore\; t = T - w$

Now, by the definition of a Laplace transform

$$\mathcal{L}\{F(t)\} = \int_0^\infty e^{-st}.F(t).dt \qquad (s > 0)$$

Since $t = T - w \qquad \therefore\; dt \equiv dT \quad$ and $\quad F(t) = F(T)$.

Also when $t = 0$, $\qquad T =$ ....

and when $\quad t = \infty$, $\qquad T =$ ....

4

$t = 0,$	$T = w$
$t = \infty,$	$T = \infty$

The integral $\mathcal{L}\{F(t)\} = \int_0^\infty e^{-st}.F(t).dt$ $\quad\cdots\quad$ (i)

can now be written in terms of T, with appropriate changes in the limits.

$$\mathcal{L}\{F(t)\} = \int_w^\infty e^{-s(T-w)}.F(T).dT$$

$$= \int_w^\infty e^{-sT}.e^{sw}.F(T)\,dT$$

$$= e^{sw}\int_w^\infty e^{-sT}.F(T)\,dT$$

$$\therefore\ e^{-sw}\,\mathcal{L}\{F(t)\} = \int_w^\infty e^{-sT}.F(T).dT$$

$$= \int_w^\infty e^{-st}.F(t).dt \qquad \cdots \quad \text{(ii)}$$

since the value of the definite integral depends on the limits to be substituted and not on the symbol chosen for the variable.

Now subtract (ii) from (i) – and what do we get?

5

$$(1 - e^{-ws})\,\mathcal{L}\{F(t)\} = \int_0^\infty e^{-st}.F(t).dt - \int_w^\infty e^{-st}.F(t).dt$$

The right-hand side = $\displaystyle\int_0^w e^{-st}.F(t).dt$

$$\therefore\ \mathcal{L}\{F(t)\} = \frac{1}{1 - e^{-ws}} \int_0^w e^{-st}.F(t).dt$$

where w = period.

This is an important result. Make a note of it in your record book.

6

$$\mathcal{L}\{F(t)\} = \frac{1}{1 - e^{-ws}} \int_0^w e^{-st}.F(t).dt$$

Note that the integral of $e^{-st}F(t)$ is taken over one cycle, i.e. from $t = 0$ to $t = w$, and not from $t = 0$ to $t = \infty$ as in our previous work.

Let us consider a simple example.

Example 1. Find the Laplace transform of the function represented by the square wave shown.

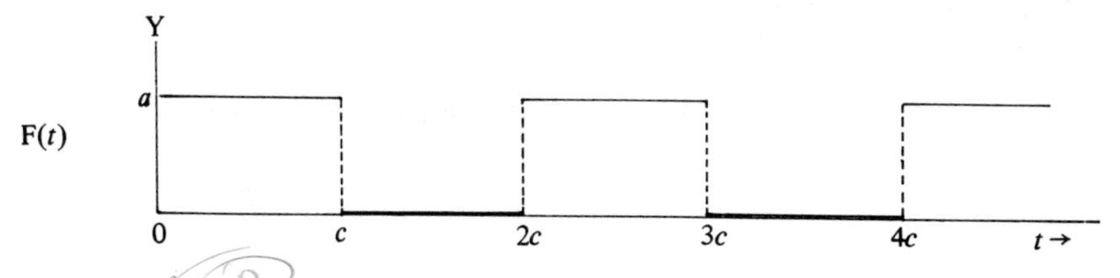

In this case: $w = 2c$, $\quad F(t) = a$ for $0 < t < c$; $\quad = 0$ for $c < t < 2c$ $\quad F(t + 2c) = F(t)$

$\therefore\ \mathcal{L}\{F(t)\} =$ ....

(as an integral. Do not evaluate it yet.)

7

$$\mathcal{L}\{F(t)\} = \frac{1}{1 - e^{-2cs}} \int_0^{2c} e^{-st}.F(t).dt$$

The function F(t) consists of two parts –

$$F(t) = a \quad \text{for} \quad 0 < t < c$$
$$F(t) = 0 \quad \text{for} \quad c < t < 2c$$

$$\therefore \mathcal{L}\{F(t)\} = \frac{1}{1 - e^{-2cs}} \left\{ \int_0^c e^{-st}.a.dt + \int_c^{2c} e^{-st}.0.dt \right\}$$

$$= \dots\dots\dots\dots\dots\dots\dots\dots\dots\dots\dots\dots$$

Finish it off.

8

$$\mathcal{L}\{F(t)\} = \frac{a}{s(1 + e^{-cs})}$$

for:

$$\mathcal{L}\{F(t)\} = \frac{a}{1 - e^{-2cs}} \left[\frac{e^{-st}}{-s} \right]_0^c$$

$$= \frac{a}{1 - e^{-2cs}} \left[\left(\frac{e^{-cs}}{-s} \right) - \left(-\frac{1}{s} \right) \right]$$

$$= \frac{a}{1 - e^{-2cs}} \left[\frac{1 - e^{-cs}}{s} \right]$$

$$\therefore \mathcal{L}\{F(t)\} = \frac{a}{s(1 + e^{-cs})}$$

It is just as easy as that. Now this one:–

Example 2. Find the Laplace transform of the function shown below –

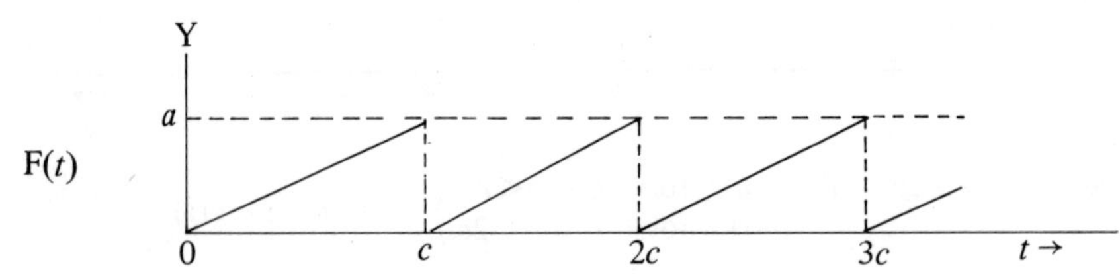

First we see that between $t = 0$ and $t = c$

F(t) = ...

9

$$\boxed{F(t) = \frac{a}{c}t}$$

i.e. a straight line through the origin, with slope $\frac{a}{c}$.

$\therefore$ For $0 < t < c$, $\quad F(t) = \frac{a}{c}t \quad$ and $\quad F(t + c) = F(t)$

$$\therefore \mathcal{L}\{F(t)\} = \frac{1}{1 - e^{-cs}} \int_0^c e^{-st} . F(t) . dt$$

It saves writing if we multiply across by $(1 - e^{-cs})$ for the time being.

$$(1 - e^{-cs})\ \mathcal{L}\{F(t)\} = \frac{a}{c} \int_0^c e^{-st} . t . dt$$

$$= \ldots\ldots\ldots\ldots\ldots\ldots\ldots\ldots\ldots\ldots\ldots\ldots$$

Evaluate the integral and then divide across by $(1 - e^{-cs})$ to obtain $\mathcal{L}\{F(t)\}$.

10

$$\boxed{\mathcal{L}\{F(t)\} = \frac{a}{cs^2}\left\{1 - \frac{cs}{e^{cs} - 1}\right\}}$$

for:

$$(1 - e^{-cs})\ \mathcal{L}\{F(t)\} = \frac{a}{c}\left\{\left[t\left(\frac{e^{-st}}{-s}\right)\right]_0^c + \frac{1}{s}\int_0^c e^{-st}\, dt\right\}$$

$$= \frac{a}{c}\left\{\left(-\frac{ce^{-sc}}{c}\right) - \left(0\right) + \frac{1}{s}\left[\frac{e^{-st}}{-s}\right]_0^c\right\}$$

$$= \frac{a}{c}\left\{-\frac{ce^{-cs}}{s} - \frac{e^{-cs}}{s^2} + \frac{1}{s^2}\right\}$$

$$= \frac{a}{cs^2}\left\{1 - e^{-cs} - cse^{-cs}\right\}$$

$$\therefore \mathcal{L}\{F(t)\} = \frac{a}{cs^2}\left\{1 - \frac{cse^{-cs}}{1 - e^{-cs}}\right\}$$

$$\therefore \mathcal{L}\{F(t)\} = \frac{a}{cs^2}\left\{1 - \frac{cs}{e^{cs} - 1}\right\}$$

Check your working. If your result looks somewhat different, it may well be correct, but expressed in a different form.

Now on to frame 11.

11

Here is another example.

Example 3. Let us consider the half-wave rectifier wave-form shown below and defined by

$$\left.\begin{aligned} F(t) &= a \sin t \qquad & 0 < t < \pi \\ &= 0 & \pi < t < 2\pi \end{aligned}\right\} \qquad F(t + 2\pi) = F(t).$$

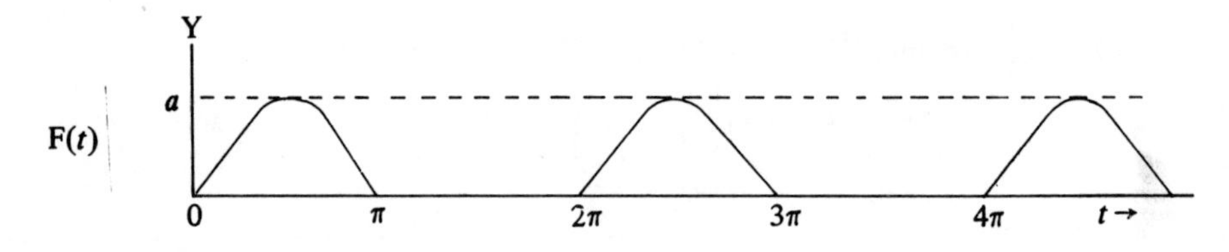

In this case, the period $= w = 2\pi$.

Then $\mathcal{L}\{F(t)\} = \dfrac{1}{1 - e^{-2\pi s}} \cdot \displaystyle\int_0^{2\pi} e^{-st} . F(t) . dt$

$\therefore\ (1 - e^{-2\pi s})\, \mathcal{L}\{F(t)\} = \ldots\ldots\ldots\ldots\ldots\ldots\ldots\ldots\ldots\ldots$

12

$$(1 - e^{-2\pi s})\, \mathcal{L}\{F(t)\} = \boxed{\int_0^{\pi} e^{-st}\, a \sin t . dt + \int_{\pi}^{2\pi} e^{-st} . 0 . dt}$$

$$= a . \mathcal{I} \int_0^{\pi} e^{-st} . e^{jt} . dt$$

since $e^{j\theta} = \cos\theta + j \sin\theta \quad \therefore\ \sin t = \mathcal{I}\{e^{jt}\}$

$$= a . \mathcal{I} \int_0^{\pi} e^{-(s-j)t}\, dt$$

$= \ldots\ldots\ldots\ldots\ldots\ldots\ldots\ldots\ldots\ldots$

Evaluate the integral; then divide across by $(1 - e^{-2\pi s})$ and so obtain $\mathcal{L}\{F(t)\}$.

Complete the job on your own.

$$\mathcal{L}\{F(t)\} = \frac{a}{(s^2+1)(1-e^{-\pi s})}$$

Here is the working:

$$(1-e^{-2\pi s})\,\mathcal{L}\{F(t)\} = a\,.\,\mathcal{I}\int_0^{\pi} e^{-(s-j)t}\,dt$$

$$= a\,.\,\mathcal{I}\left[\frac{e^{-(s-j)t}}{-(s-j)}\right]_0^{\pi}$$

$$= a\,.\,\mathcal{I}\left[-\frac{1}{s-j}\left\{e^{-(s-j)\pi}-1\right\}\right]$$

$$= a\,.\,\mathcal{I}\left[\frac{1}{s-j}\left\{1-e^{-s\pi}\,e^{j\pi}\right\}\right]$$

Now $e^{j\pi} = \cos\pi + j\sin\pi$
$= -1 + j0$

$$= a\,.\,\mathcal{I}\left[\frac{s+j}{s^2+1}\left\{1+e^{-\pi s}\right\}\right]$$

$$= a\left[\frac{1+e^{-\pi s}}{s^2+1}\right]$$

$$\therefore\ \mathcal{L}\{F(t)\} = \frac{a}{s^2+1}\cdot\frac{1+e^{-\pi s}}{1-e^{-2\pi s}} = \frac{a}{s^2+1}\cdot\frac{1}{1-e^{-\pi s}}$$

$$\mathcal{L}\{F(t)\} = \frac{a}{(s^2+1)(1-e^{-\pi s})}$$

14 As you see, there is no great difficulty, for they are all tackled in much the same way. Here is a further example.

Example 4. Sketch the function $\begin{cases} F(t) = e^t & \text{for} \quad 0 < t < 2\pi \\ F(t+2\pi) = F(t) \end{cases}$

and determine its Laplace transform.

First of all then, sketch the graph of the function between, say, $t = 0$ and $t = 6\pi$.

Then turn to frame 15.

15

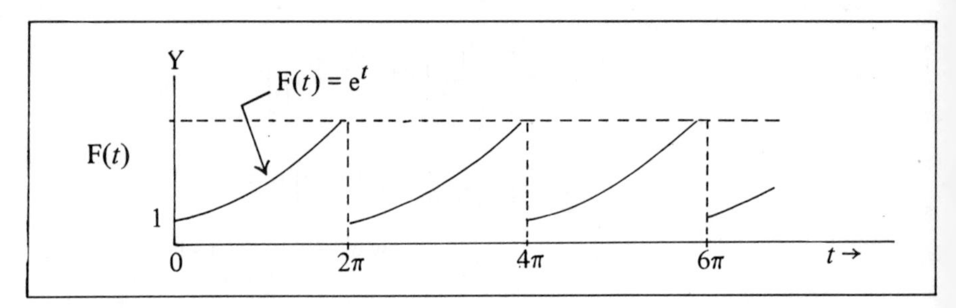

Period $= 2\pi$ i.e. $w = 2\pi$.

$\therefore \mathcal{L}\{F(t)\} = \ldots\ldots\ldots\ldots\ldots\ldots\ldots\ldots$ (In integral form)

16

$$\mathcal{L}\{F(t)\} = \frac{1}{1 - e^{-2\pi s}} \int_0^{2\pi} e^{-st} . e^t . dt$$

$\therefore$ We have

$$(1 - e^{-2\pi s})\, \mathcal{L}\{F(t)\} = \int_0^{2\pi} e^{-st} . e^t . dt$$

$$= \int_0^{2\pi} e^{-(s-1)t}\, dt$$

$$= \ldots\ldots\ldots\ldots\ldots\ldots$$

Do the next step.

17

$$(1 - e^{-2\pi s})\,\mathcal{L}\{F(t)\} = \left[\frac{e^{-(s-1)t}}{-(s-1)}\right]_0^{2\pi}$$

$\therefore\ \mathcal{L}\{F(t)\} = \ldots\ldots\ldots\ldots\ldots\ldots$

Finish it off.

18

$$\frac{1 - e^{-2(s-1)\pi}}{(s-1)(1 - e^{-2\pi s})}$$

for

$$(1 - e^{-2\pi s})\,\mathcal{L}\{F(t)\} = \left[\frac{e^{-(s-1)2\pi}}{-(s-1)}\right] - \left[\frac{1}{-(s-1)}\right]$$

$$= \frac{1 - e^{-2(s-1)\pi}}{s-1}$$

Finally, divide by the factor $(1 - e^{-2\pi s})$ on the left-hand side and we have

$$\mathcal{L}\{F(t)\} = \frac{1 - e^{-2(s-1)\pi}}{(s-1)(1 - e^{-2\pi s})}$$

So it is all very straight forward. Here is one for you to do.

Example 5. Sketch the graph of the function

$$\left.\begin{aligned} F(t) &= 1 \quad &&\text{for} \quad 0 < t < k \\ &= -1 \quad &&\text{for} \quad k < t < 2k \end{aligned}\right\} \qquad F(t + 2k) = F(t)$$

and determine the Laplace transform.

First of all, sketch the graph. Then on to the next frame.

19

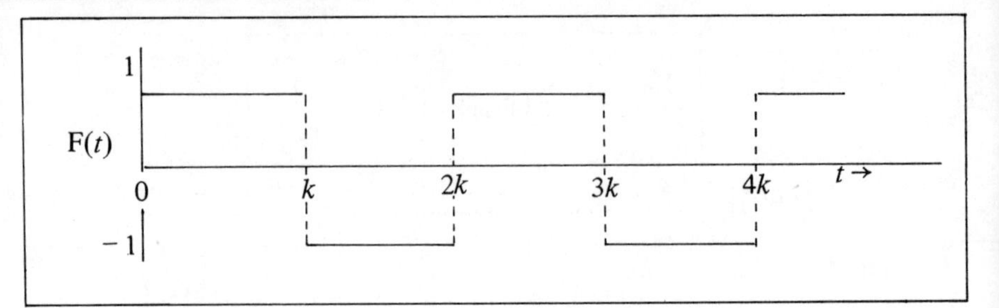

The function is clearly periodic with period $2k$. So on now to find

$$\mathcal{L}\{F(t)\} = \ldots\ldots\ldots\ldots\ldots\ldots$$

Complete the working and then check with frame 20.

20

$$\mathcal{L}\{F(t)\} = \frac{1 - e^{-ks}}{s(1 + e^{-ks})}$$

Here are the details:

$$\mathcal{L}\{F(t)\} = \frac{1}{1 - e^{-2ks}} \int_0^{2k} e^{-st} . F(t) . dt$$

$$\therefore \; (1 - e^{-2ks})\,\mathcal{L}\{F(t)\} = \int_0^k e^{-st} . 1 . dt + \int_k^{2k} e^{-st} . (-1) . dt$$

$$= \left[\frac{e^{-st}}{-s}\right]_0^k - \left[\frac{e^{-st}}{-s}\right]_k^{2k} = \frac{e^{-ks} - 1}{-s} - \frac{e^{-2ks} - e^{-ks}}{-s}$$

$$= \frac{e^{-ks} - 1 - e^{-2ks} + e^{-ks}}{-s} = \frac{1 - 2e^{-ks} + e^{-2ks}}{s} = \frac{(1 - e^{-ks})^2}{s}$$

$$\therefore \; \mathcal{L}\{F(t)\} = \frac{(1 - e^{-ks})^2}{s(1 - e^{-2ks})} = \frac{(1 - e^{-ks})^2}{s(1 - e^{-ks})(1 + e^{-ks})}$$

$$\therefore \; \mathcal{L}\{F(t)\} = \frac{1 - e^{-ks}}{s(1 + e^{-ks})}$$

Now move on to the next frame.

Inverse transforms

21

In earlier programmes, inverse transforms have been of standard forms which we have recognized, or which we have obtained by reference to a suitable table of transforms.

In the field of periodic functions, this approach is not so readily available to us, but the inverse transforms are still required. How we can obtain them, we will now see.

22

Example 6. Determine the inverse transform $\mathcal{L}^{-1}\left\{\dfrac{1-3e^{-2s}+2e^{-3s}}{s(1-e^{-3s})}\right\}$.

Sketch the graph and define it in function terms.

Before we rush ahead, let us take a look at the given transform. First of all, we recognize the factor $(1-e^{-3s})$ in the denominator as very possibly indicating a periodic function – or being derived from some similarly shaped factor. Unfortunately, the rest of the expression cannot be fitted to any of our standard transforms, since with periodic functions, $e^{-st}\,F(t)$ is integrated over one cycle instead of from $t=0$ to $t=\infty$.

How then, do we deal with a problem such as this?

The general hint is to write $(1-e^{-3s})$ in the denominator as $(1-e^{-3s})^{-1}$ in the numerator and to expand it as a binomial series, i.e. like $(1-x)^{-1}$.

$(1-x)^{-1}=$ ....

23

$$(1-x)^{-1}=1+x+x^2+x^3+\ldots$$

$\therefore\ (1-e^{-3s})^{-1}=$ ....

24

$$(1-e^{-3s})^{-1}=1+e^{-3s}+e^{-6s}+e^{-9s}+e^{-12s}+\ldots$$

$$\therefore\ \mathcal{L}\{F(t)\}=\frac{1-3e^{-2s}+2e^{-3s}}{s(1-e^{-3s})}=\frac{(1-3e^{-2s}+2e^{-3s})(1-e^{-3s})^{-1}}{s}$$

$$=\frac{1}{s}(1-3e^{-2s}+2e^{-3s})(1+e^{-3s}+e^{-6s}+e^{-9s}+\ldots)$$

Multiplying the second bracket by each term of the first bracket in turn and collecting up like terms, we get

$\mathcal{L}\{F(t)\}=$ ....

25

$$\frac{1}{s}\left\{1 - 3e^{-2s} + 3e^{-3s} - 3e^{-5s} + 2e^{-6s} + \dots\right\}$$

for:

$$\mathcal{L}\{F(t)\} = \frac{1}{s}\left\{1 + e^{-3s} + e^{-6s} + e^{-9s} + \dots - 3e^{-2s} - 3e^{-5s} - 3e^{-8s} - 3e^{-11s} \dots + 2e^{-3s} + 2e^{-6s} + 2e^{-9s} + \dots\right\}$$

$$= \frac{1}{s}\left\{1 - 3e^{-2s} + 3e^{-3s} - 3e^{-5s} + 3e^{-6s} - 3e^{-8s} + 3e^{-9s} - \dots\right\}$$

Now from the previous programme, you no doubt remember that

$$\mathcal{L}^{-1}\left\{\frac{e^{-cs}}{s}\right\} = \dots\dots\dots\dots$$

26

$$\mathcal{L}^{-1}\left\{\frac{e^{-cs}}{s}\right\} = H(t-c)$$

$\therefore$ Since $\mathcal{L}\{F(t)\} = \frac{1}{s}\left\{1 - 3e^{-2s} + 3e^{-3s} - 3e^{-5s} + 3e^{-6s} - 3e^{-8s} \dots\right\}$

then $\quad F(t) = \dots\dots\dots\dots$

27

$$F(t) = H(t) - 3\,H(t-2) + 3\,H(t-3) - 3\,H(t-5) + \dots$$

So we are now back to the case of a periodic function expressed as an infinite series in terms of the Heaviside unit step function.

Using the result above, sketch the graph of the function

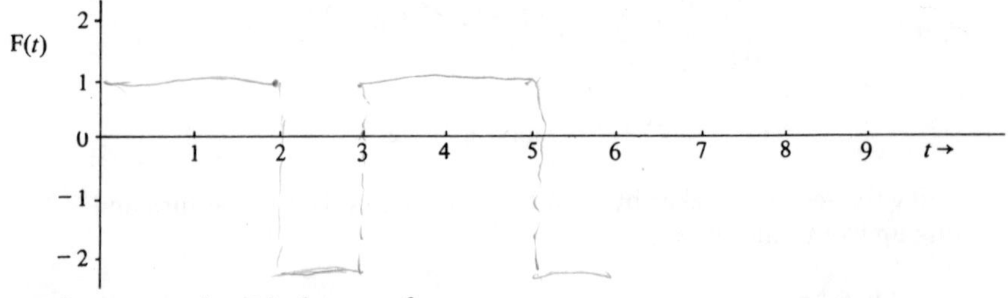

Check your result with the next frame.

28

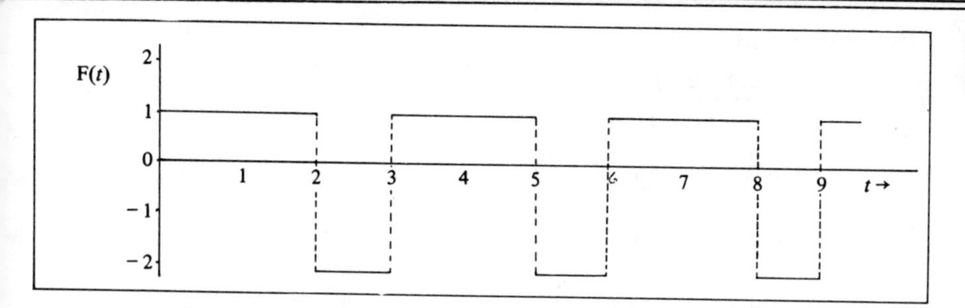

So (i) we have expressed the result in terms of the Heaviside unit function,
(ii) we have sketched the graph of the function,
and (iii) we now have to state the result in function form. That should not cause any difficulty.

$F(t) =$ .... for ....

$=$ .... for ....

and .... $=$ ....

29

$F(t) = 1$	for	$0 < t < 2$
$= -2$	for	$2 < t < 3$
$F(t + 3) = F(t)$		

So although we have the special method for finding the transforms of periodic functions, when we come to find inverse transforms, we usually have to proceed via the Heaviside unit step functions.

Now here is one for you to work through.

Example 7. Determine the inverse transform $\mathcal{L}^{-1}\left\{\dfrac{2(1 - e^{-s})}{s(1 - e^{-4s})}\right\}$

Sketch the waveform and express the result in function form.

Work right through it to the end and then check your results by reference to the next frame, where the working is set out in detail.

30

Here is the solution:

(i) $\mathcal{L}\{F(t)\} = \dfrac{2(1-e^{-s})}{s(1-e^{-4s})}$

$$= \frac{2}{s}.(1-e^{-s})(1-e^{-4s})^{-1}$$

$$= \frac{2}{s}.(1-e^{-s})(1+e^{-4s}+e^{-8s}+e^{-12s}+e^{-16s}+\ldots)$$

$$= \frac{2}{s}\left\{\begin{matrix} 1 & +e^{-4s} & +e^{-8s} & +e^{-12s} & \ldots \\ -e^{-s} & -e^{-5s} & -e^{-9s} & & \ldots \end{matrix}\right\}.$$

$$= \frac{2}{s}\left\{1-e^{-s}+e^{-4s}-e^{-5s}+e^{-8s}-e^{-9s}\ldots\right\}$$

$$\therefore \mathcal{L}\{F(t)\} = \frac{2}{s} - \frac{2e^{-s}}{s} + \frac{2e^{-4s}}{s} - \frac{2e^{-5s}}{s} + \frac{2e^{-8s}}{s} - \ldots$$

$$\therefore \underline{F(t) = 2\,H(t) - 2\,H(t-1) + 2\,H(t-4) - 2\,H(t-5) + 2\,H(t-8)\ldots}$$

(ii) Sketch graph:

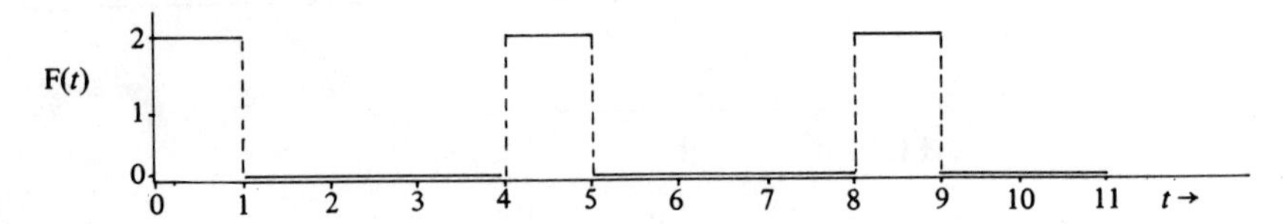

(iii) Function form:

$$\begin{cases} F(t) = 2 & \text{for} \quad 0<t<1 \\ \quad\;\; = 0 & \text{for} \quad 1<t<4 \\ F(t+4) = F(t) & \end{cases}$$

And that is it! Do you agree?.

All that now remains is the Test Exercise. First, look through the programme again and be sure that you fully understand all we have done, including the worked examples.

Then, when you are ready, turn on to the next frame.

Answer all the questions. There are no tricks.

31

Test Exercise–VI

1. Determine the Laplace transforms of the periodic functions having the following waveforms:

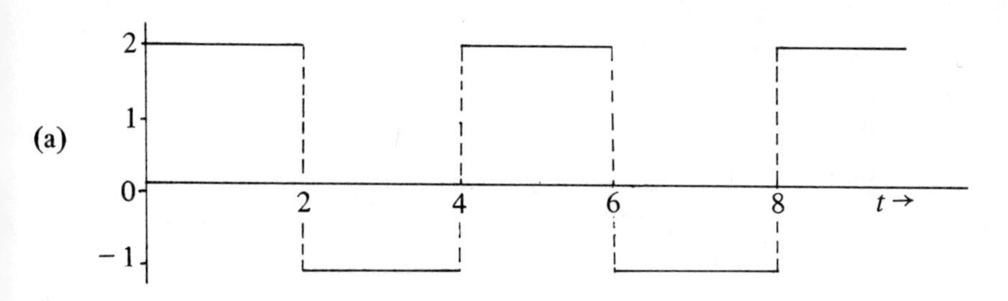

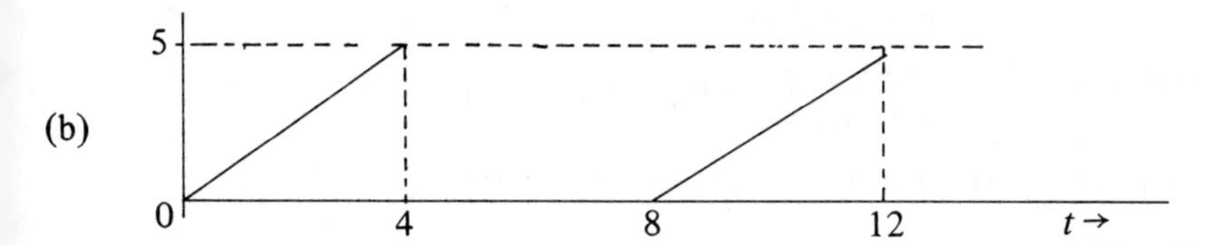

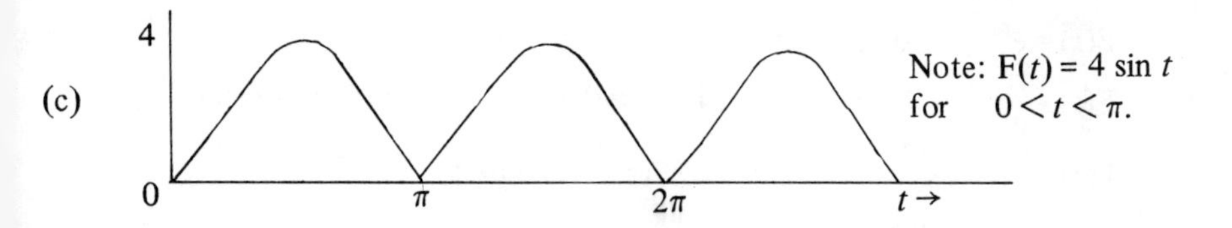

2. Sketch the graph of the function

$$\begin{cases} F(t) = 2\,e^{-t} & \text{for} \quad 0 < t < 5 \\ F(t+5) = F(t) \end{cases}$$

between $t = 0$ and $t = 10$, and hence determine its Laplace transform.

3. Find the inverse transform

$$\mathcal{L}^{-1}\left\{\frac{2(1 + 2e^{-2s} - 3e^{-3s})}{s(1 - e^{-3s})}\right\}$$

Sketch the graph and state the result in function form.

Further Problems – VI

Find the Laplace transform of the periodic functions defined in questions 1 to 13.

1. $F(t) = t^2 \quad 0<t<2$; $= 4 \quad 2<t<3$; $\quad F(t+3) = F(t)$
2. $F(t) = e^{2t} \quad 0<t<1$; $= 1 \quad 1<t<2$; $\quad F(t+2) = F(t)$
3. $F(t) = t \quad 0<t<1$; $= 0 \quad 1<t<2$; $\quad F(t+2) = F(t)$
4. $F(t) = \sin t \quad 0<t<\pi$; $= 0 \quad \pi<t<2\pi$; $\quad F(t+2\pi) = F(t)$
5. $F(t) = t \quad 0<t<2\pi \quad F(t+2\pi) = F(t)$
6. $F(t) = E \quad 0<t<\pi/w$; $= -E \quad \pi/w<t<2\pi/w$; $\quad F(t+2\pi/w) = F(t)$
7. $F(t) = 1 \quad 0<t<a$; $= -1 \quad a<t<2a$; $\quad F(t+2na) = F(t)$
8. $F(t) = t \quad 0<t<W \quad F(t+w) = F(t)$
9. $F(t) = t \quad 0<t<\pi$; $= 0 \quad \pi<t<2\pi$; $\quad F(t+2\pi) = F(t)$
10. $F(t) = e^t \quad 0<t<2\pi \quad F(t+2\pi) = F(t)$
11. $F(t) = \dfrac{t}{a} \quad 0<t<a \quad F(t+na) = F(t)$ (n = positive integer)
12. $F(t) = t^2 \quad o<t<2\pi \quad F(t+2\pi) = F(t)$
13. $F(t) = a\left|\sin \dfrac{2\pi t}{T}\right|$ for $0<t<T$
14. If $F(t) = c$ for $0<t<\dfrac{T}{2}$ $\quad = 0$ for $\dfrac{T}{2}<t<T$

    show that $\mathcal{L}\{F(t)\} = \dfrac{c}{s(1+e^{-sT/2})}$

    Hence solve $\ddot{y} + w^2 y = F(t)$ where $F(t)$ is defined above, given that at $t = 0$, $y = 0, \dot{y} = 0$. Find the value of y at $t = T$.
15. Determine $\mathcal{L}\{F(t)\}$ given that

    $F(t) = e^{-t} \quad 0<t<1 \quad F(t+1) = F(t)$

    If $\dot{y} + 2y = F(t)$, as defined above, solve the equation for y in terms of t, for $2<t<3$, given that at $t = 0, y = 0$.

Programme 7

DIRAC DELTA FUNCTION (IMPULSE FUNCTION)

1

Introduction

In the first four programmes of this series, we dealt with the definition of Laplace transforms and their general use in solving *differential equations,* and we worked through numerous examples of this kind.

We then gave attention to the *Heaviside unit step function*, which provided us with the means of denoting the switching on and off of items in the physical problem – and there are many applications of this.

We now consider the *impulse function* representing, for example, an extremely large force acting for a minutely small interval of time, or a 'shock' voltage applied to an electrical circuit. It is equivalent to a sudden blow, rather than a continuous applied force. This is quite different from anything we have met in this work, so let us first of all consider this new function.

2

Consider a single rectangular pulse of width b and height $1/b$ centred at $t = a$.

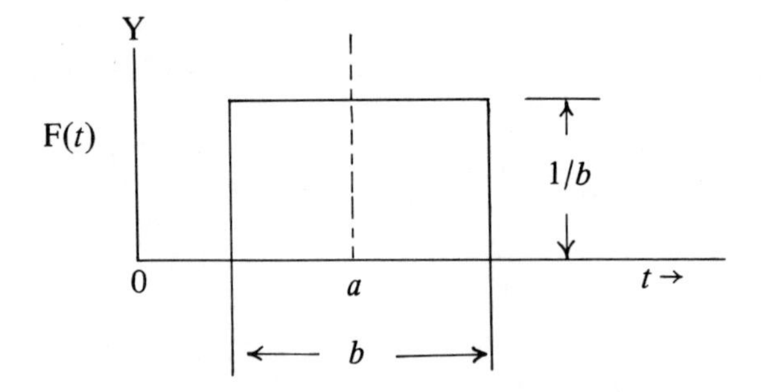

Note that by definition, the area of this pulse $= b.\dfrac{1}{b} = 1.$

If the width of the pulse is reduced to half its original value and the area remains 1 unit of area, sketch on a single diagram both the original form and the new form of the pulse.

When you have done that, move on to the next frame.

3

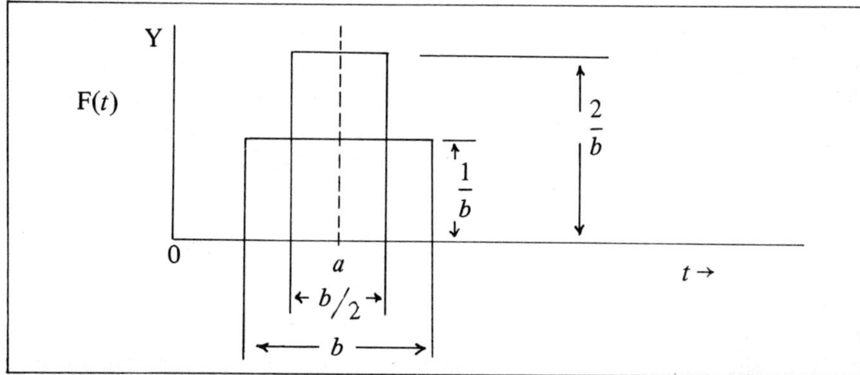

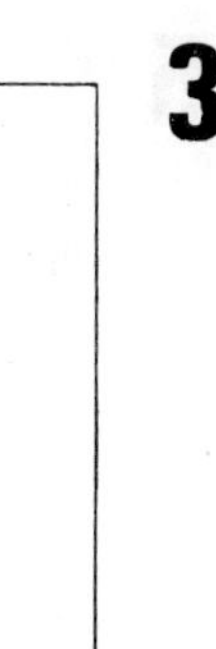

Clearly, if the width is halved, the height must be doubled if the area of the pulse is to remain as unity.

If this process is continued, keeping the area = 1, then as the width $b \to 0$, the height $\frac{1}{b} \to$ ∞ ....

4

∞

We thus approach the final stage, where the function exists for a minutely small time interval, but during that minute interval, the magnitude of the function is extremely large.

The function represented by the limiting stage of this process is called the *Dirac delta function* or the *unit impulse function* and is denoted by $\delta(t-a)$.

Graphically, it would need a rectangular pulse of zero width and infinite height and, for convenience, is represented by a single vertical line surmounted by an arrow head, thus –

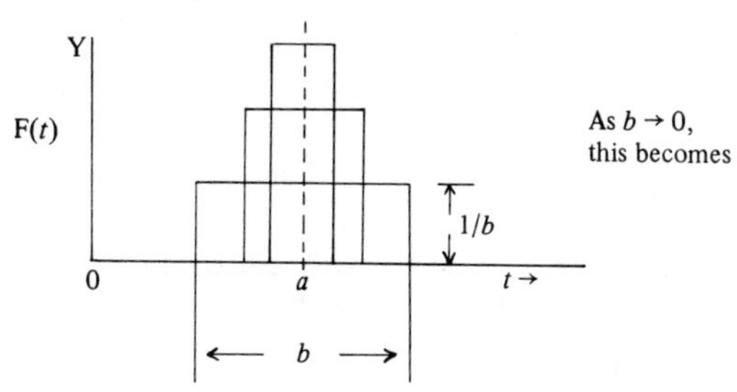

As $b \to 0$, this becomes

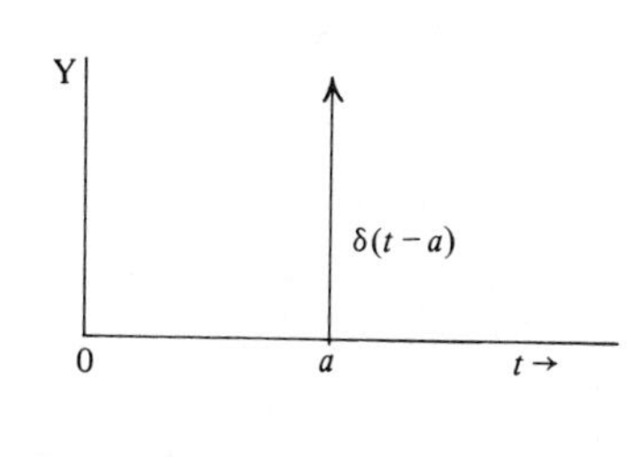

At all stages, and in the limiting state, the area of the pulse is 1.

....

5

1

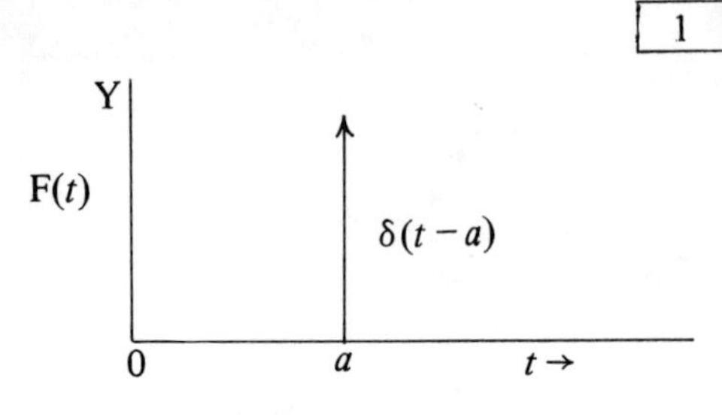

Note that the delta function $\delta(t-a)$ exists only at $t = a$, at which it is infinitely great.

i.e.
$$\begin{aligned}\delta(t-a) &= 0 \qquad t \neq a\\ &= \infty \qquad t = a\end{aligned}$$

The impulse function at the origin

This is merely a special case of $\delta(t-a)$ in which $a = 0$. Hence the function is denoted by $\delta(t)$.

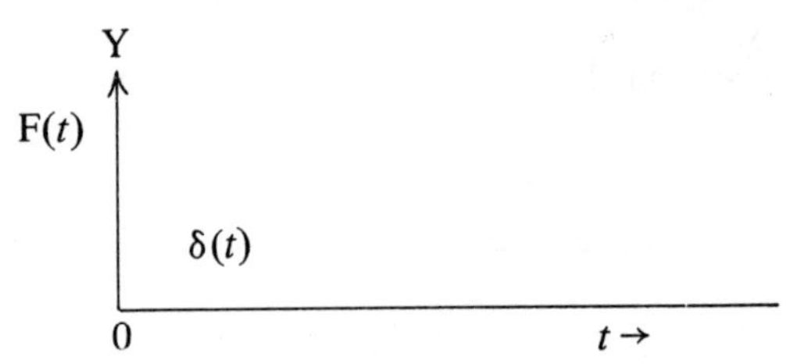

The Dirac delta function $\delta(t-a)$ is thus a special function and we must investigate some of its properties.

On then to frame 6.

6

(1) If we integrate the function $\delta(t-a)$ between $t = p$ and $t = q$, where $t = a$ is between $t = p$ and $t = q$,

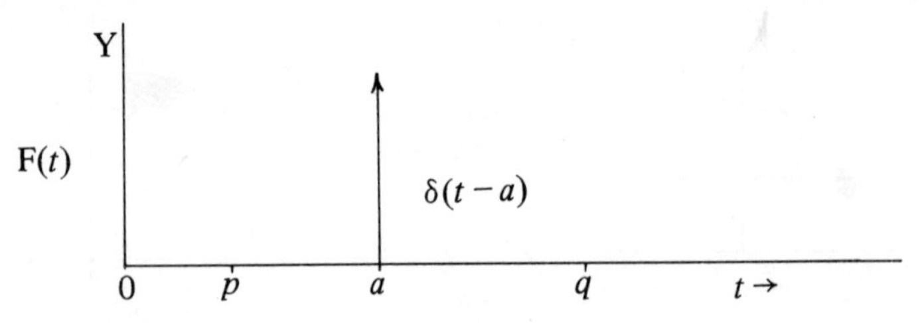

then remembering the definition of $\delta(t-a)$,

$$\int_p^q \delta(t-a).\,dt = \dots\dots\dots\dots$$

7

1

since:

(i) for $p < t < a$, $\delta(t-a) = 0$ $\quad\therefore\quad \int_p^a \delta(t-a).dt = 0.$

(ii) at $t = a$, area of the pulse = 1, by definition.

(iii) for $a < t < q$, $\delta(t-a) = 0$ $\quad\therefore\quad \int_a^q \delta(t-a).dt = 0.$

$\therefore$ The total area between $t = p$ and $t = q$, $= 0 + 1 + 0 = 1$ unit of area.

$$\therefore \int_p^q \delta(t-a)\,dt = 1. \quad \text{provided} \quad p < t < q \qquad \ldots \text{I}$$

Make a note of this result – and then move on.

8

(2) Let us now consider

$\int_p^q F(t).\delta(-a).dt$ for $p < a < q$, as before, where $F(t)$ is a given function of t.

The integrand, i.e. the product $F(t).\delta(t-a)$ is zero for all values of t within the range $t = p$ to $t = q$, except at the point $t = a$.

At $t = a$ (or over a minute interval around $t = a$), the function $F(t)$ may be regarded as constant, having the value $F(a)$ and this constant factor may be taken outside the integral.

$\therefore$ For $p < a < q$,

$$\int_p^q F(t).\delta(t-a).dt = F(a) \int_p^q \delta(t-a).dt$$

$$= \ldots\ldots\ldots\ldots\ldots\ldots$$

9

$$\boxed{F(a)}$$

i.e.
$$\int_p^q F(t).\delta(t-a).dt = F(a)\int_p^q \delta(t-a).dt$$
$$= F(a).1$$
$$= F(a).$$

$$\therefore \int_p^q F(t).\delta(t-a).dt = F(a) \qquad \ldots \text{ II}$$

This is also important, so make a note of this result.

$$\therefore \int_2^5 (t^2-6).\delta(t-4)\,dt = [t^2-6]_{t=4}$$
$$= 16-6 = \underline{10}$$

and
$$\int_0^{\pi} \sin 3t.\delta\left(t-\frac{\pi}{2}\right)dt = \ldots\ldots$$

10

$$\boxed{-1}$$

since
$$\int_0^{\pi} \sin 3t.\delta\left(t-\frac{\pi}{2}\right).dt = [\sin 3t]_{t=\pi/2}$$
$$= \sin\frac{3\pi}{2}$$
$$= \underline{-1}$$

Likewise:

(i) $\int_0^4 5.\delta(t-2).dt = \ldots\ldots$

(ii) $\int_1^6 e^{-3t}.\delta(t-3).dt = \ldots\ldots$

(iii) $\int_0^{\infty} (2t^2-5t+8).\delta(t-1)\,dt = \ldots\ldots$

11

(i) $\displaystyle\int_0^4 5\,.\,\delta(t-2)\,.\,dt = 5\,.\,1 = \underline{5}$

(ii) $\displaystyle\int_1^6 e^{-3t}\,.\,\delta(t-3)\,.\,dt = [e^{-3t}]_{t=3} = \underline{e^{-9}}$

(iii) $\displaystyle\int_0^\infty (2t^2-5t+8)\,.\,\delta(t-1)\,dt = [2t^2-5t+8]_{t=1}$

$= 2 - 5 + 8 = \underline{5}$

And, in general, if $p < a < q$, then

$$\int_p^q F(t)\,.\,\delta(t-a)\,.\,dt = \ldots\ldots\ldots\ldots\ldots\ldots$$

12

$$\boxed{F(a)}$$

(3) *Laplace transform of* $\delta(t-a)$

We have established that

$$\int_p^q F(t)\,.\,\delta(t-a)\,.\,dt = F(a) \qquad p < a < q$$

This also applies, of course, when $p = 0$ and $q = \infty$.

$$\int_0^\infty F(t)\,.\,\delta(t-a)\,.\,dt = F(a) \qquad a > 0 \quad \text{i.e.} \quad 0 < a < \infty.$$

Putting $F(t) = e^{-st}$, this becomes

$$\int_0^\infty e^{-st}\,.\,\delta(t-a)\,dt = \mathcal{L}\{\delta(t-a)\}$$

$$= \ldots\ldots\ldots\ldots\ldots\ldots$$

13

$$\boxed{e^{-as}}$$

i.e. the value of F(t) at $t = a$

i.e. the value of e^{-st} at $t = a$, i.e. e^{-as}

$$\underline{\mathcal{L}\{\delta(t-a)\} = e^{-as}} \qquad \ldots \text{ III}$$

Add this to the list. Then continue.

It follows directly from result III that the Laplace transform of the impulse function at the origin is ....

14

$$\boxed{1}$$

since for $\delta(t)$, $a = 0$. $\therefore e^{-as} = e^0 = 1$

$$\therefore \underline{\mathcal{L}\{\delta(t)\} = 1}$$

This is the only function whose Laplace transform is 1.

15

(4) Consider now $\mathcal{L}\{F(t) \cdot \delta(t-a)\}$

$$\mathcal{L}\{F(t) \cdot \delta(t-a)\} = \int_0^\infty e^{-st} \cdot F(t) \cdot \delta(t-a) \cdot dt$$

Now $e^{-st} F(t) \cdot \delta(t-a) = 0$ for all values of t, except at $t = a$

when $e^{-st} = e^{-as}$

and $F(t) = F(a)$

$$\therefore \mathcal{L}\{F(t) \cdot \delta(t-a)\} = F(a) \cdot e^{-as} \int_0^\infty \delta(t-a)\, dt$$

$$= F(a)\, e^{-as} \cdot 1$$

$$\therefore \underline{\mathcal{L}\{F(t) \cdot \delta(t-a)\} = F(a)\, e^{-as}} \qquad \ldots \text{ IV}$$

Note this result and then turn on to frame 16.

16

So:

(i) $\mathcal{L}\{5 \,.\, \delta(t-3)\} = \ldots\ldots$

(ii) $\mathcal{L}\{t^2 \,.\, \delta(t-2)\} = \ldots\ldots$

(iii) $\mathcal{L}\{\cos 2t \,.\, \delta(t-\pi)\} = \ldots\ldots$

(iv) $\mathcal{L}\{\sinh t \,.\, \delta(t)\} = \ldots\ldots$

17

(i) $\mathcal{L}\{5 \,.\, \delta(t-3)\} = \underline{5\,\mathrm{e}^{-3s}}$

(ii) $\mathcal{L}\{t^2 \,.\, \delta(t-2)\} = \underline{4 \,.\, \mathrm{e}^{-2s}}$

(iii) $\mathcal{L}\{\cos 2t \,.\, \delta(t-\pi)\} = \cos 2\pi \,.\, \mathrm{e}^{-\pi s} = \underline{\mathrm{e}^{-\pi s}}$

(iv) $\mathcal{L}\{\sinh t \,.\, \delta(t)\} = \sinh 0 \,.\, \mathrm{e}^{0} = 0.1 = \underline{0}$

So far, in our results, we have established a number of important conclusions. Complete the following without looking at your notes

(i) $\int_p^q \delta(t-a) \,.\, dt = \ldots\ldots$ provided $\ldots\ldots$

(ii) $\int_p^q \mathrm{F}(t) \,.\, \delta(t-a) \,.\, dt = \ldots\ldots$ provided $\ldots\ldots$

(iii) $\mathcal{L}\{\delta(t-a)\} = \ldots\ldots$

(iv) $\mathcal{L}\{\delta(t)\} = \ldots\ldots$

(v) $\mathcal{L}\{\mathrm{F}(t) \,.\, \delta(t-a)\} = \ldots\ldots$

Check your answers with the next frame.

18

(i) $\int_p^q \delta(t-a) \, . \, dt = \underline{1}$ provided $\underline{p < a < q.}$

(ii) $\int_p^q F(t) \, . \, \delta(t-a) \, dt = \underline{F(a).}$ provided $\underline{p < a < q.}$

(iii) $\mathcal{L}\{\delta(t-a)\} = \underline{e^{-as}}$

(iv) $\mathcal{L}\{\delta(t)\} = \underline{1}$

(v) $\mathcal{L}\{F(t) \, . \, \delta(t-a)\} = \underline{F(a) \, . \, e^{-as}}$

These are the main tools when dealing with the Dirac delta function.

Move on now to frame 19.

19 *Example.* Impulses of 1, 2, 5 units occur at $t = c$, $t = 2c$, and $t = 3c$ respectively, as shown.

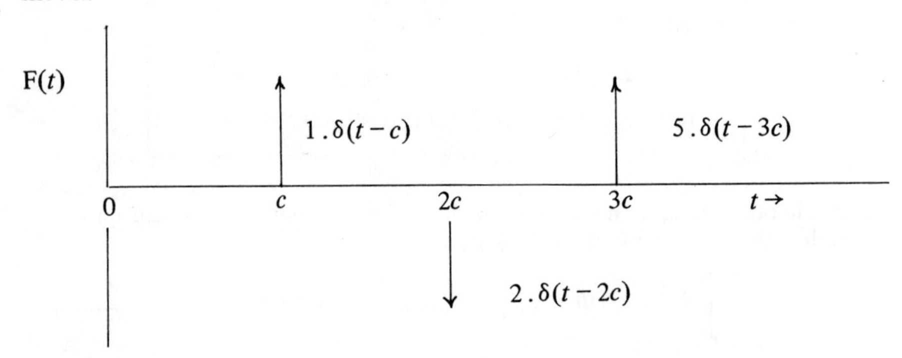

Write down an expression for F(t) and hence find its Laplace transform.

20

$$F(t) = 1 \, . \, \delta(t-c) - 2 \, . \, \delta(t-2c) + 5 \, . \, \delta(t-3c)$$

$$\therefore \mathcal{L}\{F(t)\} = e^{-cs} - 2e^{-2cs} + 5e^{-3cs}$$

It is just as easy as that.

So on now to the next piece of work.

21

(5) A further interesting result can be obtained from the fact that the area of the pulse, by definition, is unity,

$$\text{i.e.} \quad \int_0^t \delta(t-a)\,dt = 0 \qquad \text{for} \qquad t<a$$

$$= 1 \qquad \text{for} \qquad t>a$$

The right-hand side is the definition of the Heaviside unit step function since

$$\mathrm{H}(t-a) = 0 \qquad \text{for} \qquad t<a$$

$$= 1 \qquad \text{for} \qquad t>a$$

$$\therefore \quad \int_0^t \delta(t-a)\,.\,dt = \mathrm{H}(t-a)$$

$$\therefore \ \delta(t-a) = \frac{d}{dt}\{\mathrm{H}(t-a)\} \qquad \ldots \quad \mathrm{V}$$

and it follows

$$\delta(t) = \frac{d}{dt}\{\mathrm{H}(t)\}$$

There is then a close connection between the Dirac delta function and the Heaviside unit step function – which is not really surprising since we developed the delta function from a single rectangular pulse which could itself be expressed in terms of the unit step function.

Now for one or two examples, so on to the next frame.

22

Example 1. An impulsive voltage $E.\delta(t)$ is applied to a circuit containing inductance L and capacitance C in series. If current and charge are initially zero, find an expression for the current at time t.

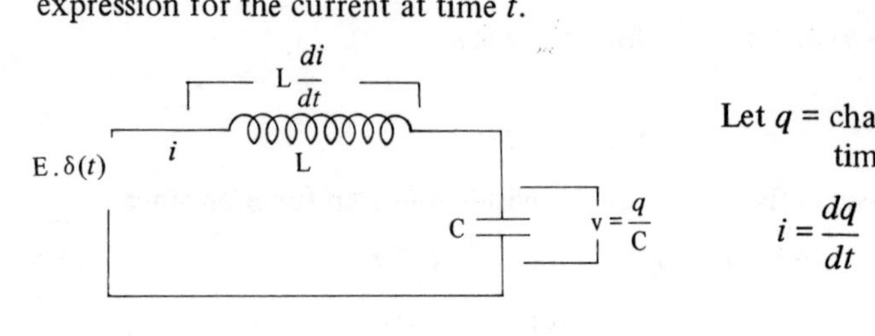

Let q = charge on the capacitor at time t.

$$i = \frac{dq}{dt}$$

Voltage across the inductor $= L\dfrac{di}{dt}$: voltage across the capacitor $= \dfrac{q}{C}$. Since the total voltage drop is equal to the applied voltage, the mathematical model of this problem is the differential equation

....

23

$$L\frac{di}{dt} + \frac{q}{C} = E.\delta(t)$$

To solve this we proceed through the usual steps.

(i) Express the equation in Laplace transforms (remembering that q is also a variable).

....

24

$$L(s\bar{i} - i_0) + \frac{1}{C}.\bar{q} = E.1$$

since $\mathcal{L}\{\delta(t)\} = 1$. Now we have to replace $\bar{q}$ in terms of $\bar{i}$.

We know that $i = \dfrac{dq}{dt}$. $\quad\therefore \bar{i} = s\bar{q} - q_0$.

From the information given

i_0 = ...

q_0 = ...

25

$$i_0 = 0; \qquad q_0 = 0$$

$$\therefore \bar{i} = s\bar{q} \qquad \therefore \bar{q} = \frac{\bar{i}}{s}$$

∴ The equation becomes

....

26

$$\boxed{Ls\bar{i} + \frac{\bar{i}}{Cs} = E}$$

$$\bar{i}\left\{Ls + \frac{1}{Cs}\right\} = E$$

$\therefore \bar{i} =$ ....

27

$$\boxed{\bar{i} = \frac{Es}{Ls^2 + \frac{1}{C}}}$$

for: $\bar{i}\left\{Ls^2 + \frac{1}{C}\right\} = Es \qquad \therefore \bar{i} = \frac{Es}{Ls^2 + \frac{1}{C}}$

$$\therefore \bar{i} = \frac{E}{L} \cdot \frac{s}{s^2 + \frac{1}{LC}}$$

$\therefore i =$ ....

28

$$i = \frac{E}{L} . \cos\left\{\frac{t}{\sqrt{(LC)}}\right\}$$

Here is the solution complete.

$$L\frac{di}{dt} + \frac{q}{C} = E . \delta(t) \qquad i = \frac{dq}{dt}$$

$$L(s\bar{\imath} - i_0) + \frac{\bar{q}}{C} = E . 1 \qquad \bar{\imath} = s\bar{q} - q_0$$

$$i_0 = q_0 = 0 \qquad \therefore \bar{\imath} = s\bar{q} \qquad \therefore \bar{q} = \frac{\bar{\imath}}{s}$$

$$\therefore Ls\bar{\imath} + \frac{\bar{\imath}}{Cs} = E$$

$$\therefore \bar{\imath}\left\{Ls^2 + \frac{1}{C}\right\} = Es$$

$$\therefore \bar{\imath} = \frac{Es}{Ls^2 + \frac{1}{C}} = \frac{E}{L} . \frac{s}{s^2 + \frac{1}{LC}}$$

$$\therefore i = \frac{E}{L} . \cos\left\{\frac{t}{\sqrt{(LC)}}\right\}$$

Note that when the impulse occurs at $t = 0$, the driving function is $E\,\delta(t)$ and, on taking transforms, this becomes E . 1, since $\mathcal{L}\{\delta(t)\} = 1$. The solution of the equation then is quite straightforward.

Another example in the next frame.

29

Example 2. A mass M is attached to a spring of stiffness k^2 and rests in equilibrium on a frictionless horizontal plane. The mass is subjected to an impulsive blow P at $t = 0$, to extend the spring. Find an expression for the displacement at time t.

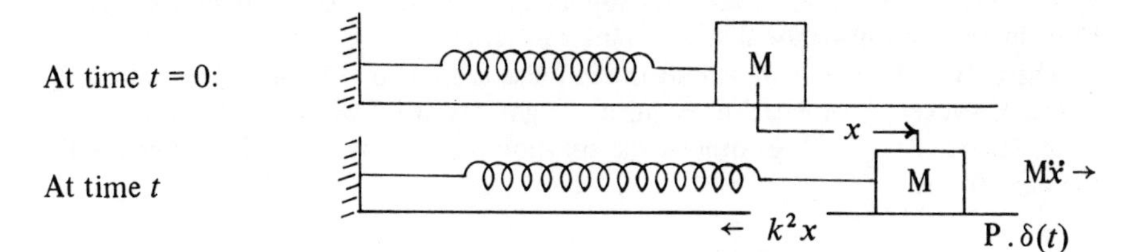

$$\text{At } t = 0, x = 0, \dot{x} = 0.$$

So the equation of motion is $M\ddot{x} = P.\delta(t) - k^2x$

$$\therefore\ M\ddot{x} + k^2x = P.\delta(t).$$

Now go ahead and solve this equation and so determine

$x = \ldots\ldots\ldots\ldots\ldots\ldots\ldots\ldots\ldots\ldots\ldots\ldots\ldots\ldots\ldots\ldots$

30

$$\boxed{x = \frac{P}{k\sqrt{M}}\ \sin\left\{\frac{kt}{\sqrt{M}}\right\}}$$

For, we have:

$$M(s^2\bar{x} - sx_0 - x_1) + k^2\bar{x} = P.1$$

$$x_0 = x_1 = 0 \qquad \therefore\ Ms^2\bar{x} + k^2\bar{x} = P$$

$$\therefore\ (Ms^2 + k^2)\bar{x} = P$$

$$\therefore\ \bar{x} = \frac{P}{Ms^2 + k^2} = \frac{P}{M}\cdot\frac{1}{s^2 + \dfrac{k^2}{M}}$$

$$\therefore\ x = \frac{P}{M}\cdot\frac{\sqrt{M}}{k}\cdot \sin\left\{\frac{kt}{\sqrt{M}}\right\}$$

$$\therefore\ x = \underline{\frac{P}{k\sqrt{M}}\cdot \sin\left\{\frac{kt}{\sqrt{M}}\right\}}$$

Now for another.

31

Example 3. A system has the equation of motion

$$\ddot{x} + 5\dot{x} + 6x = \mathrm{F}(t)$$

where $\mathrm{F}(t)$ is an impulse of 10 units applied at $t = 4$. At $t = 0, x = 0$, and $\dot{x} = 2$. Find an expression for the displacement x in terms of t.

The only difference here is that the impulse is applied at $t = 4$ and not at the origin. However, the treatment is quite straightforward if we remember that $\mathcal{L}\{\mathrm{P}\,.\,\delta(t-a)\} = \mathrm{P}\,.\,\mathrm{e}^{-as}$. Complete the solution on your own and then check with the next frame.

32

$$\boxed{x = 2\,\mathrm{e}^{-2t}\left\{1 + 5\,.\,\mathrm{e}^{8}\,.\,\mathrm{H}(t-4)\right\} - 2\mathrm{e}^{-3t}\left\{1 + 5\,.\,\mathrm{e}^{12}\,.\,\mathrm{H}(t-4)\right\}}$$

Maybe you have the result in a slightly different form. Anyway, check the working:

$$\ddot{x} + 5\dot{x} + 6x = 10\,.\,\delta(t-4) \qquad t = 0, x = 0, \dot{x} = 2.$$

$$\therefore\ (s^2\bar{x} - sx_0 - x_1) + 5(s\bar{x} - x_0) + 6\bar{x} = 10\,\mathrm{e}^{-4s} \qquad x_0 = 0, x_1 = 2.$$

$$\therefore\ s^2x - 2 + 5s\bar{x} + 6\bar{x} = 10\,\mathrm{e}^{-4s}$$

$$\therefore\ (s^2 + 5s + 6)\bar{x} = 2 + 10\,\mathrm{e}^{-4s}$$

$$\therefore\ \bar{x} = (2 + 10\,\mathrm{e}^{-4s})\,\frac{1}{(s+2)(s+3)}$$

$$\bar{x} = (2 + 10\,\mathrm{e}^{-4s})\left\{\frac{1}{s+2} - \frac{1}{s+3}\right\}$$

$$\bar{x} = 2\left\{\frac{1}{s+2} - \frac{1}{s+3}\right\} + 10\left\{\frac{\mathrm{e}^{-4s}}{s+2} - \frac{\mathrm{e}^{-4s}}{s+3}\right\}$$

$$\therefore\ x = 2\left\{\mathrm{e}^{-2t} - \mathrm{e}^{-3t}\right\} + 10\left\{\mathrm{e}^{-2(t-4)}\,\mathrm{H}(t-4) - \mathrm{e}^{-3(t-4)}\,\mathrm{H}(t-4)\right\}.$$

$$= 2\left\{\mathrm{e}^{-2t} - \mathrm{e}^{-3t}\right\} + 10\left\{\mathrm{e}^{-2t}\,\mathrm{e}^{8}\,\mathrm{H}(t-4) - \mathrm{e}^{-3t}\,\mathrm{e}^{12}\,\mathrm{H}(t-4)\right\}.$$

$$\therefore\ \underline{x = 2\,\mathrm{e}^{-2t}\left\{1 + 5\,.\,\mathrm{e}^{8}\,.\,\mathrm{H}(t-4)\right\} - 2\,\mathrm{e}^{-3t}\left\{1 + 5\,.\,\mathrm{e}^{12}\,.\,\mathrm{H}(t-4)\right\}}$$

And so on to frame 33.

33

This brings us to the end of the present programme.

Before you work through the Test Exercise, read down the Summary list set out below which collects together the main results that we have established. If there are any points on which you are not completely clear, go back to that part of the programme and work through it again. It is important to be sure that you understand.

Summary

1. Delta function: $\delta(t-a) = 0 \quad t \neq a$
   $\qquad = \infty \quad t = a$

2. Delta function at the origin: $\delta(t) = 0 \quad t \neq 0$
   $\qquad = \infty \quad t = 0$

3. Area of pulse = 1

$$\int_p^q \delta(t-a)\,.\,dt = 1 \qquad p < a < q$$

4. $$\int_p^q \mathrm{F}(t)\,.\,\delta(t-a)\,.\,dt = \mathrm{F}(a) \qquad p < a < q$$

5. $\mathcal{L}\{\delta(t-a)\} = \mathrm{e}^{-as}$

   $\mathcal{L}\{\delta(t)\} = 1$

6. $\mathcal{L}\{\mathrm{F}(t)\,.\,\delta(t-a)\} = \mathrm{F}(a)\,.\,\mathrm{e}^{-as}$

Now turn on to the Test Exercise.

34

Here is the usual exercise covering the work you have covered in this programme. All the questions are similar to those you have tackled before, so there will be no difficulty. Take your time over the test.

Test Exercise–VII

Answer all the questions

1. Evaluate

(i) $\int_0^5 e^{-4t}.\delta(t-2).dt$

(ii) $\int_0^\infty 2\cos 3t.\delta(t-\pi)\,dt$

(iii) $\int_1^5 (3t^2+4).\delta(t-3)\,dt.$

2. Determine

(i) $\mathcal{L}\{5.\delta(t-2)\}$

(ii) $\mathcal{L}\{\sin 5t.\delta(t-\pi/2)\}$

(iii) $\mathcal{L}\{e^{-2t}.\delta(t-1)\}$

3. Sketch the graph of

$F(t) = 2.\delta(t) + 3.\delta(t-2) - 4.\delta(t-5)$

and determine $\mathcal{L}\{F(t)\}$.

4. Solve the equation

$\ddot{x} + 2\dot{x} + 5x = 5.\delta(t)$

given that at $t = 0, x = 3, \dot{x} = 0$.

5. The equation of motion of a system is

$\ddot{x} + 4\dot{x} + 3x = 2.\delta(t-6)$

At $t = 0, x = 0, \dot{x} = 2$. Find an expression for x in terms of t.

Further Problems – VII

1. Find $\mathcal{L}\{F(t)\}$ where $F(t) = t\,.\,H(t-1) + t^2\,.\,\delta(t-1)$.

2. Solve the equations

$$3\dot{x} + 4\dot{y} + x - 12y = 2\,.\,\delta(t)$$
$$2\dot{x} + 3\dot{y} + x - 6y = 0$$

At $t = 0, x = 1, y = 2$.

3. A mass M is attached to a spring of stiffness w^2M and set in motion at $t = 0$ by an impulse P. The equation of motion is

$$M\ddot{x} + Mw^2x = P\,.\,\delta(t).$$

Solve the equation to obtain an expression for x.

4. An impulsive voltage E is applied at $t = 0$ to a circuit containing inductance L and capacitance C in series. Initially, current and charge are zero. Find an expression for the current i in the circuit at time t.

5. A voltage $E\,.\,\delta(t)$ is applied to a series L, R, C circuit with zero initial conditions. Show that $L\dfrac{di}{dt} + Ri + \dfrac{q}{C} = E\,.\,\delta(t)$ and hence determine i if $CR^2 < 4L$.

6. A mass m moves in a straight line, so that when at a distance x from a fixed point 0 in the line, it is subjected to a restoring force mw^2x. At $t = 0$, m is at rest. At $t = 0$ and at $t = \dfrac{\pi}{2w}$, equal impulses I are applied at the mass, both in the positive direction of x. Find the displacement at time t.

7. A light beam of length l is clamped horizontally at each end. A concentrated load W is applied at $x = a$, where x is the distance from one end. The resulting deflection y at any point is given by

$$EI\frac{d^4y}{dx^4} = W\,.\,\delta(x-a)$$

with $y = \dfrac{dy}{dx} = 0$ at $x = 0$ and at $x = l$. Show that

$$y = \frac{W}{6EI}\left\{(x-a)^3\,H(x-a) + \frac{3a(l-a)^2x^2}{l^2} - \frac{(l-a)^2(l+2a)x^3}{l^3}\right\}$$

8. An e.m.f. E is applied to a series circuit containing inductance L, resistance R and capacitance C, where $R^2/(4L^2) > 1/(LC)$. If the current and charge are initially zero and E is an impulsive voltage applied at $t = 0$, determine an expression for the voltage across the capacitor at time t.

Programme 8

USEFUL THEOREMS INCLUDING THE CONVOLUTION THEOREM

1

1. Laplace transform of $\int_0^t F(t)\,.dt$

In electrical circuit problems, we often need to express the voltage developed across a capacitor at time t seconds. This depends on the capacitance of the capacitor and on the charge present at the particular instant.

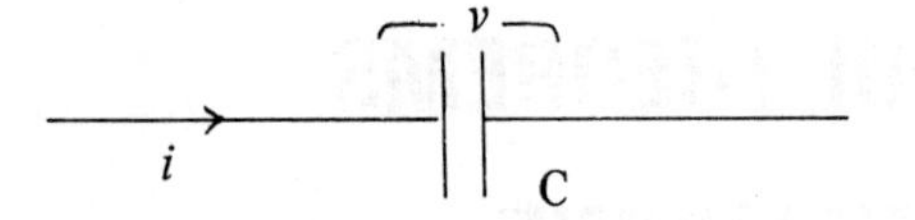

If q is the charge, then $v = \dfrac{q}{C}$

While the capacitance C is fixed and determined by the design of the capacitor, the charge is a variable and is the sum of the current flowing into the capacitor.

i.e. $q = \int_0^t i\,dt \qquad \therefore\ i =$ ....

2

$$\boxed{i = \frac{dq}{dt}}$$

We have used this statement on a number of occasions in the past, and on taking transforms, this becomes

$\bar{i} =$ ....

3

$$\boxed{\bar{i} = s\bar{q} - q_0}$$

If the initial conditions are zero, then $q_0 = 0$ and $i = sq$

$\therefore\ \bar{q} =$ ....

$$\boxed{\bar{q} = \frac{\bar{i}}{s}.}$$

Thus, if $\frac{\bar{i}}{s}$ is substituted for $\bar{q}$ in the particular differential equation, the solution can be obtained in the normal manner.

For example, for a series circuit containing R and C, with applied voltage E and with $i = q = 0$ at $t = 0$, we have:

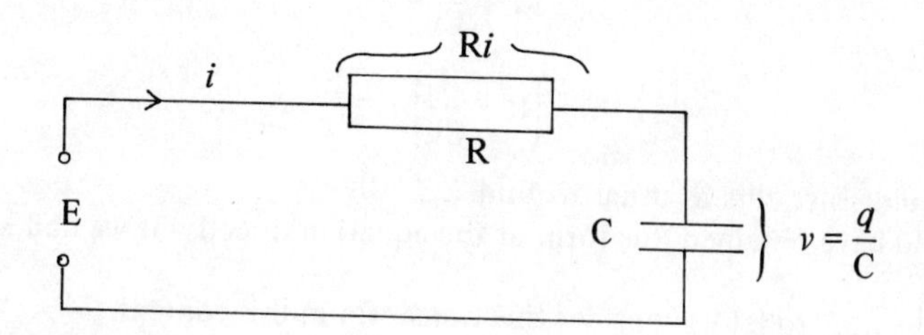

$\therefore$ The equation for i is

$$\mathrm{R}i + v = \mathrm{E}$$

i.e.
$$\mathrm{R}i + \frac{q}{\mathrm{C}} = \mathrm{E}$$

Expressing this in transforms, the equation becomes

....

5

$$R\bar{i} + \frac{\bar{q}}{C} = \frac{E}{s}$$

Now $i = \dfrac{dq}{dt}$ $\quad\therefore \bar{i} = s\bar{q} - q_0$

At $t = 0, q = 0$ $\quad\therefore q_0 = 0$ $\quad\therefore \bar{i} = s\bar{q}$ $\quad\therefore \bar{q} = \dfrac{1}{s}\bar{i}$

So the equation can now be written

$$R\bar{i} + \frac{\bar{i}}{Cs} = \frac{E}{s}$$

$$\therefore \quad \left\{R + \frac{1}{Cs}\right\}\bar{i} = \frac{E}{s}$$

and we could solve this as usual to find i.

We could have obtained this form of the equation directly, if we had known $\mathcal{L}\left\{\int_0^t i\,dt\right\}$ . . . . . so let us consider this transform in this context.

On then to the next frame.

6 We had $\quad q = \int_0^t i\,dt \quad \therefore \bar{q} = \mathcal{L}\left\{\int_0^t i\,dt\right\}$ . . . (i)

Also $\quad i = \dfrac{dq}{dt} \quad \therefore \bar{i} = s\bar{q} - q_0$

$$\therefore \bar{q} = \frac{1}{s}(\bar{i} + q_0) \quad \text{. . . (ii)}$$

$\therefore$ We have

$$\bar{q} = \mathcal{L}\left\{\int_0^t i\,dt\right\}$$

and

$$\bar{q} = \frac{1}{s}(\bar{i} + q_0)$$

$$\underline{\mathcal{L}\left\{\int_0^t i\,dt\right\} = \frac{1}{s}(\bar{i} + q_0)}$$

and, for the case of zero initial conditions, this becomes

....

7

$$\mathcal{L}\left\{\int_0^t i\,dt\right\} = \frac{1}{s}\cdot\bar{\imath}$$

The problem we were considering can now be tackled thus:

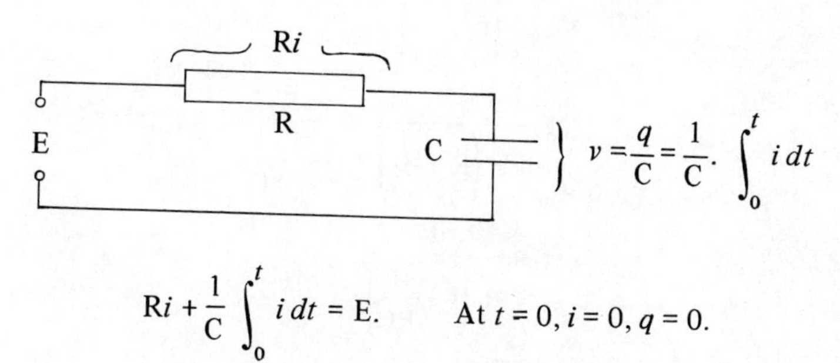

$$v = \frac{q}{C} = \frac{1}{C}\cdot\int_0^t i\,dt$$

$$Ri + \frac{1}{C}\int_0^t i\,dt = E. \qquad \text{At } t = 0, i = 0, q = 0.$$

Expressing this in transforms, with zero initial conditions,

....

8

$$R\bar{\imath} + \frac{1}{C}\cdot\frac{1}{s}\bar{\imath} = \frac{E}{s}$$

i.e.

$$\left\{R + \frac{1}{Cs}\right\}\bar{\imath} = \frac{E}{s}$$

and we have arrived at the same equation, but rather more quickly.

Finish off the solution, giving

$i =$ ....

9

$$\boxed{i = \frac{E}{R} \cdot e^{-\frac{t}{RC}}}$$

for

$$\left\{R + \frac{1}{Cs}\right\}\bar{\imath} = \frac{E}{s}$$

$$\therefore \bar{\imath} = \frac{E}{s\left\{R + \frac{1}{Cs}\right\}} = \frac{E}{sR + \frac{1}{C}}$$

$$= \frac{E}{R\left\{s + \frac{1}{RC}\right\}}$$

$$\therefore \bar{\imath} = \frac{E}{R} \cdot \frac{1}{s + \frac{1}{RC}}$$

$$\therefore i = \frac{E}{R} \cdot e^{-\frac{t}{RC}}$$

So although not essential, it is convenient to know that

$$\mathcal{L}\left\{\int_0^t i\,dt\right\} = \ldots\ldots\ldots\ldots\ldots\ldots\ldots\ldots\ldots\ldots\ldots\ldots\ldots\ldots\ldots$$

10

$$\boxed{\mathcal{L}\left\{\int_0^t i\,dt\right\} = \frac{1}{s}(\bar{\imath} + q_0)}$$

where q_0 is the charge at $t = 0$.

Make a note of this result: it is well worth remembering.

Then on to frame 11.

11

In general, if $y = \int_0^t \mathrm{F}(t)\,dt$ then $\mathrm{F}(t) = \dfrac{dy}{dt}$.

$$\therefore f(s) = \mathcal{L}\{\mathrm{F}(t)\} = s\bar{y} - y_0$$

$$\therefore s\bar{y} = f(s) + y_0$$

$$\bar{y} = \frac{1}{s}\left\{f(s) + y_0\right\}$$

$$\therefore \mathcal{L}\left\{\int_0^t \mathrm{F}(t)\,dt\right\} = \frac{1}{s}\left\{f(s) + y_0\right\} \qquad \ldots \text{ I}$$

For zero initial conditions

$$\mathcal{L}\left\{\int_0^t \mathrm{F}(t)\,dt\right\} = \frac{1}{s}.\,f(s)$$

Copy these down into your record book for future reference.

Example. If $y = \int_0^t (t^2 + 2\cos t)\,dt$ and $y = 0$ at $t = 0$,

$\mathcal{L}\{y\} =$ ....

12

$$\boxed{\frac{2}{s^4} + \frac{2}{s^2+1}}$$

for

$$f(s) = \frac{2}{s^3} + \frac{2s}{s^2+1}$$

$$\therefore \mathcal{L}\{y\} = \frac{1}{s}.\,f(s) = \frac{2}{s^4} + \frac{2}{s^2+1}$$

13

2. Heaviside expansion theorem

To find inverse transforms, we often express $f(s)$ in partial fractions, by applying the various methods we have covered.

Let $f(s) = \dfrac{P(s)}{Q(s)}$ where $Q(s)$ is a polynomial of degree n and $P(s)$ is a polynomial of lower degree.

Expressed in partial fractions, this has the form

$$f(s) = \frac{P(s)}{Q(s)} = \frac{A_1}{s - a_1} + \frac{A_2}{s - a_2} + \ldots \frac{A_k}{s - a_k} + \ldots \frac{A_n}{s - a_n}.$$

To find A_k we multiply both sides by $(s - a_k)$

$$\therefore\ (s - a_k)\frac{P(s)}{Q(s)} = \frac{A_1(s - a_k)}{s - a_1} + \frac{A_2(s - a_k)}{s - a_2} + \ldots + A_k + \ldots$$

$$\text{Then } A_k = \lim_{s \to a_k}\left\{(s - a_k)\frac{P(s)}{Q(s)}\right\}$$

$$= \lim_{s \to a_k}\left\{P(s) . \frac{(s - a_k)}{Q(s)}\right\}$$

Since $(s - a_k)$ is a factor of $Q(s)$, this gives $\dfrac{0}{0}$ (= ?)

By l'Hôpital's rule

$$A_k = P(a_k) . \lim_{s \to a_k}\left\{\frac{1}{Q'(s)}\right\} \qquad \text{where} \qquad Q'(s) = \frac{d}{ds}\left\{Q(s)\right\}$$

$$\therefore\ A_k = \frac{P(a_k)}{Q'(a_k)}$$

$$\therefore\ f(s) = \frac{P(s)}{Q(s)} = \frac{P(a_1)}{Q'(a_1)} . \frac{1}{s - a_1} + \frac{P(a_2)}{Q'(a_2)} . \frac{1}{s - a_2} + \ldots + \frac{P(a_k)}{Q'(a_k)} . \frac{1}{s - a_k} + \ldots$$

$$\therefore\ F(t) = \frac{P(a_1)}{Q'(a_1)} e^{a_1 t} + \frac{P(a_2)}{Q'(a_2)} e^{a_2 t} + \ldots + \frac{P(a_k)}{Q'(a_k)} e^{a_k t} + \ldots \qquad \ldots \text{ II}$$

Make note of this: then turn on to frame 14 for an example.

14

This expansion theorem can be applied only if the denominator $Q(s)$ is expressed as a product of linear factors $(s - a_k)$.

Example 1. Find $F(t)$ if $f(s) = \dfrac{s^2 + 4}{(s-1)(s-2)(s+3)}$

$$\text{Now } f(s) = \frac{P(s)}{Q(s)} \text{ and in this case}$$

$$P(s) = s^2 + 4$$

$$Q(s) = (s-1)(s-2)(s+3) = (s-1)(s^2 + s - 6)$$

$$= s^3 - 7s + 6$$

$$\therefore Q'(s) = 3s^2 - 7$$

$$P(s) = s^2 + 4; \qquad Q'(s) = 3s^2 - 7$$

Also: $a_1 = 1, a_2 = 2, a_3 = -3$

$\therefore F(t) =$ ....

Apply the expansion from frame 13.

15

$$\boxed{F(t) = -\frac{5}{4}e^{t} + \frac{8}{5}e^{2t} + \frac{13}{20}e^{-3t}}$$

for each term is of the form $\dfrac{P(a)}{Q'(a)} . e^{at}$ where $a = 1, 2, -3$.

Here is another example.

Example 2. Find $\mathcal{L}^{-1}\left\{\dfrac{2s-3}{(s-2)(s^2+4)}\right\}$

Remember that we can apply this method only if the denominator $Q(s)$ is expressed in linear factors, so what about the factor $(s^2 + 4)$? This we write as $(s + j2)(s - j2)$ and then apply the routine as before.

In this case

$$P(s) = 2s - 3$$

$Q(s) =$ ....

$Q'(s) =$ ....

16

$$Q(s) = s^3 - 2s^2 + 4s - 8$$
$$Q'(s) = 3s^2 - 4s + 4$$

So that $$\frac{P(s)}{Q'(s)} = \frac{2s-3}{3s^2 - 4s + 4}$$

Also, since $Q(s) = (s-2)(s+j2)(s-j2)$

$a_1 = \ldots\ldots\ldots\ldots$; $a_2 = \ldots\ldots\ldots\ldots$; $a_3 = \ldots\ldots\ldots\ldots$

17

$$a_1 = 2; \quad a_2 = -j2; \quad a_3 = j2$$

$\therefore\ F(t) = \ldots\ldots\ldots\ldots\ldots\ldots\ldots\ldots$

18

$$F(t) = \frac{1}{8}\{e^{2t} + 7\sin 2t - \cos 2t\}$$

Working:

$$F(t) = \frac{1}{8}e^{2t} + \frac{3+j4}{8-j8}e^{-j2t} + \frac{3-j4}{8+j8}e^{j2t}$$

$$= \frac{1}{8}\left\{e^{2t} + \frac{3+j4}{1-j}e^{-j2t} + \frac{3-j4}{1+j}e^{j2t}\right\}$$

$$= \frac{1}{8}\left\{e^{2t} + \frac{(1+j)(3+j4)}{2}e^{-j2t} + \frac{(1-j)(3-j4)}{2}e^{j2t}\right\}$$

$$= \frac{1}{8}\left\{e^{2t} + \frac{-1+j7}{2}e^{-j2t} + \frac{-1-j7}{2}e^{j2t}\right\}$$

$$= \frac{1}{8}\left\{e^{2t} - \frac{1-j7}{2}(\cos 2t - j\sin 2t) - \frac{1+j7}{2}(\cos 2t + j\sin 2t)\right\}$$

$$= \frac{1}{8}\left\{e^{2t} - \frac{1}{2}[2\cos 2t - 14\sin 2t]\right\}$$

$$\underline{F(t) = \frac{1}{8}\left\{e^{2t} + 7\sin 2t - \cos 2t\right\}}$$

Of course, we could have reached the same conclusion by using the methods that we have already practised at length in earlier programmes. This is just another method which is sometimes convenient to use.

19

3. Heaviside series expansion

This method is useful for investigating solutions for small values of t, which are then more quickly determined than by obtaining the full solutions.

As before $f(s) = \dfrac{P(s)}{Q(s)}$ where $\begin{cases} Q(s) = \text{polynomial of degree } n \\ P(s) = \text{polynomial of lower degree.} \end{cases}$

The method now is to divide numerator and denominator of $f(s)$ by s^n and to expand in ascending powers of $\dfrac{1}{s}$ by the binomial theorem.

We thus arrive at $f(s) = \dfrac{a_1}{s} + \dfrac{a_2}{s^2} + \dfrac{a_3}{s^3} + \ldots$

and inverse transforms then give

$$F(t) = a_1 + a_2 t + \frac{a_3 t^2}{2} + \ldots \qquad \ldots \quad \text{III}$$

Provided *t is small,* the solution is given by the first few terms.

For an example, move on to the next frame.

20

Example. Find F(t) for small values of t, given that $f(s) = \dfrac{s+2}{s^2 - 2s + 1}$.

So
$$f(s) = \frac{s+2}{s^2 - 2s + 1}$$

In this case, n (degree of denominator) = 2.
$\therefore$ Divide top and bottom by s^2, which gives

$f(s) =$ ....

21

$$f(s) = \frac{\dfrac{1}{s} + \dfrac{2}{s^2}}{1 - \dfrac{2}{s} + \dfrac{1}{s^2}}$$

$$\therefore\ f(s) = \left(\frac{1}{s} + \frac{2}{s^2}\right)\left(1 - \left[\frac{2}{s} - \frac{1}{s^2}\right]\right)^{-1}$$

$$= \left(\frac{1}{s} + \frac{2}{s^2}\right)\left\{1 + \left[\frac{2}{s} - \frac{1}{s^2}\right] + \left[\frac{2}{s} - \frac{1}{s^2}\right]^2 + \ldots\right\}$$

$$= \left(\frac{1}{s} + \frac{2}{s^2}\right)\left\{1 + \frac{2}{s} - \frac{1}{s^2} + \frac{4}{s^2} - \frac{4}{s^3} + \frac{1}{s^4} + \ldots\right\}$$

$$= \left(\frac{1}{s} + \frac{2}{s^2}\right)\left\{1 + \frac{2}{s} + \frac{3}{s^2} - \frac{4}{s^3} + \ldots\right\}$$

Multiplying out and collecting up terms, we get

$f(s) =$ ....

22

$$f(s) = \frac{1}{s} + \frac{4}{s^2} + \frac{7}{s^3} + \frac{2}{s^4} + \ldots$$

$\therefore\ F(t) =$ ....

$$F(t) = 1 + 4t + \frac{7}{2}t^2 + \ldots \qquad (t \text{ small})$$

Note that the full solution obtained by the normal method is

$$F(t) = e^t(1 + 3t)$$

Expanding this in series form

$$F(t) = (1 + 3t)\left\{1 + t + \frac{t^2}{2} + \frac{t^3}{6} + \ldots\right\}$$

$$= 1 + t + \frac{t^2}{2} + \frac{t^3}{6} + \ldots + 3t + 3t^2 + \frac{3t^2}{2} + \ldots$$

$$\underline{F(t) = 1 + 4t + \frac{7}{2}t^2 + \ldots\ldots}$$

which agrees with the result obtained above.

24

4. Initial value theorem

By the Heaviside series expansion, we have

$$f(s) = \frac{a_1}{s} + \frac{a_2}{s^2} + \frac{a_3}{s^3} + \ldots$$

$$\therefore\ s \, . \, f(s) = a_1 + \frac{a_2}{s} + \frac{a_3}{s^2} + \ldots$$

If $s \to \infty$, all terms on the right-hand side disappear except a_1.

$$\therefore\ a_1 = \lim_{s \to \infty} \left\{ s \, . \, f(s) \right\} \qquad \ldots \quad \text{(i)}$$

$$= \lim_{s \to \infty} \left\{ s \frac{\mathrm{P}(s)}{\mathrm{Q}(s)} \right\}$$

Also $\quad \mathrm{F}(t) = a_1 + a_2 t + \dfrac{a_3 t^2}{2} + \ldots$

$$\therefore\ a_1 = \lim_{t \to 0} \left\{ \mathrm{F}(t) \right\} \qquad \ldots \quad \text{(ii)}$$

$$\therefore\ \lim_{t \to 0} \left\{ \mathrm{F}(t) \right\} = \lim_{s \to \infty} \left\{ s \, . \, f(s) \right\} \qquad \ldots \quad \text{IV}$$

This is the *initial value theorem.* Knowing $f(s)$ of the solution, the value of $\mathrm{F}(t)$ as $t \to 0$ is given by $\lim \left\{ s \, . \, f(s) \right\}$ as $s \to \infty$.

Example. Consider $\quad f(s) = \dfrac{s}{s^2 - 4}$

Then $\quad s \, . \, f(s) = \dfrac{s^2}{s^2 - 4} = \dfrac{1}{1 - \dfrac{4}{s^2}}$

$$\therefore\ \lim_{s \to \infty} \left\{ s \, . \, f(s) \right\} = \ldots\ldots\ldots\ldots$$

25

1

for $\quad \lim_{s \to \infty} \left\{ s \, . \, f(s) \right\} = \dfrac{1}{1 - 0} = 1.$

$$\therefore\ \lim_{t \to 0} \mathrm{F}(t) = \underline{1}$$

which is correct since $\mathcal{L}^{-1} \left\{ \dfrac{s}{s^2 - 4} \right\} = \cosh 2t \quad$ and at $t = 0$, $\cosh 2t = 1$.

Now on to the next item which is closely related to this last piece of work.

5. Final value theorem

26

This theorem may be applied to determine the behaviour of a function F(t) for large values of t.

Let $f(s) = \dfrac{P(s)}{Q(s)}$ where P(s) and Q(s) are defined as earlier.

Also, let $F'(t)$ denote $\dfrac{d}{dt}\{F(t)\}$.

$$\therefore \mathcal{L}\{F'(t)\} = \int_0^\infty e^{-st} F'(t)\,dt = s.f(s) - F(0)$$

As $s \to 0$, $$\lim_{s\to 0}\left\{\int_0^\infty e^{-st}\, F'(t)\,dt\right\} = \int_0^\infty F'(t)\,dt$$

$$= \lim_{s\to\infty} \int_0^s F'(t)\,dt$$

$$= \lim_{s\to\infty} \{F(s) - F(0)\}$$

$$= \lim_{t\to\infty} \{F(t) - F(0)\}$$

$$\therefore \lim_{s\to 0} \{s.f(s) - F(0)\} = \lim_{t\to\infty} \{F(t) - F(0)\}$$

$$\therefore \underline{\lim_{t\to\infty} \{F(t)\} = \lim_{s\to 0} \{s.f(s)\}} \qquad \ldots \quad \text{V}$$

This is the final value theorem and denotes the value to which the function F(t) approaches as $t \to \infty$.

Example. To evaluate $F(t) = t^n e^{-at}, a > 0,$ as $t \to \infty$.

$F(t) = t^n e^{-at}$ $\quad \mathcal{L}\{t^n\} = \dfrac{n!}{s^{n+1}}$ $\quad \therefore \mathcal{L}\{F(t)\} = f(s) = \dfrac{n!}{(s+a)^{n+1}}$

$$\therefore \lim_{t\to\infty} \{F(t)\} = \lim_{s\to 0} \left\{\frac{sn!}{(s+a)^{n+1}}\right\} = 0$$

$$\therefore \underline{\lim_{t\to\infty} \{F(t)\} = 0}$$

On to the next.

27

So we have these two results:

(i) Initial value theorem. $\lim_{t \to 0} \{F(t)\} = \ldots\ldots\ldots\ldots\ldots\ldots\ldots\ldots\ldots\ldots\ldots$

(ii) Final value theorem. $\lim_{t \to \infty} \{F(t)\} = \ldots\ldots\ldots\ldots\ldots\ldots\ldots\ldots\ldots\ldots\ldots$

28

$$\lim_{t \to 0} \{F(t)\} = \lim_{s \to \infty} \{s \,.\, f(s)\}$$

$$\lim_{t \to \infty} \{F(t)\} = \lim_{s \to 0} \{s \,.\, f(s)\}$$

Example.

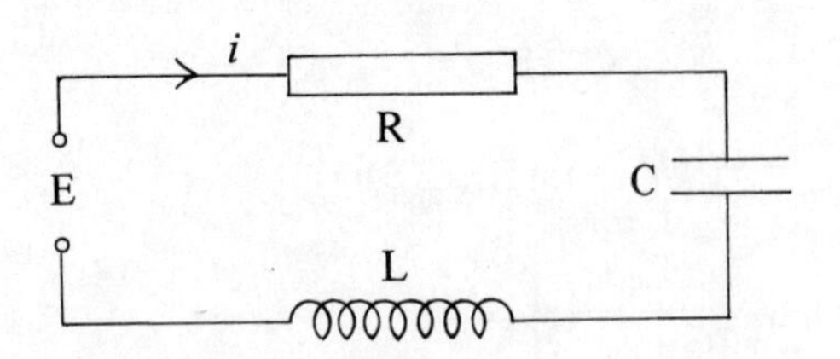

The transform of the current for the circuit shown is

$$\bar{i} = \frac{E}{L\left\{s^2 + \frac{R}{L}s + \frac{1}{LC}\right\}}$$

So, by the initial value theorem,

$\lim_{t \to 0} \{i\} = \ldots\ldots\ldots\ldots\ldots\ldots\ldots\ldots\ldots\ldots\ldots\ldots\ldots\ldots\ldots\ldots\ldots\ldots\ldots\ldots\ldots$

29

0

since

$$\lim_{t \to 0} \{F(t)\} = \lim_{s \to \infty} \{s . f(s)\} = \lim_{s \to \infty} \{s\bar{i}\}$$

$$= \lim_{s \to \infty} \left\{ \frac{Es}{L\left(s^2 + \frac{R}{L}s + \frac{1}{LC}\right)} \right\}$$

$$= \lim_{s \to \infty} \left\{ \frac{E}{L\,s + \frac{R}{L} + \frac{1}{LCs}} \right\} = \underline{0}$$

Also, by the final value theorem

$$\lim_{t \to \infty} \{i\} = \ldots\ldots\ldots\ldots\ldots\ldots\ldots$$

30

0

for

$$\lim_{t \to \infty} \{i\} = \lim_{s \to 0} \{s\bar{i}\}$$

$$= \lim_{s \to 0} \left\{ \frac{E}{L\left(s^2 + \frac{Rs}{L} + \frac{1}{LC}\right)} \right\}$$

$$= \lim_{s \to 0} \left\{ \frac{E}{L\left(s + \frac{R}{L} + \frac{1}{LCs}\right)} \right\}$$

$$= \underline{0}$$

So, in this case, both the initial and final values of i are zero.

Now let us move on to something different.

31

Inverse transforms of products of transforms

It would be very convenient to have a means of finding $F(t)$ when $f(s)$ is the product of two transforms whose inverses we recognize.

e.g. $\dfrac{s}{(s+2)(s^2+4)}$ can be regarded as $\dfrac{1}{s+2} \cdot \dfrac{s}{s^2+4}$

and we know that $\mathcal{L}^{-1}\left\{\dfrac{1}{s+2}\right\} =$ ....

and that $\mathcal{L}^{-1}\left\{\dfrac{s}{s^2+4}\right\} =$ ....

32

$$\mathcal{L}^{-1}\left\{\frac{1}{s+2}\right\} = e^{-2t}$$

$$\mathcal{L}^{-1}\left\{\frac{s}{s^2+4}\right\} = \cos 2t$$

But, of course, we cannot say that $\mathcal{L}^{-1}\left\{\dfrac{1}{s+2} \cdot \dfrac{s}{s^2+4}\right\}$ is merely the product of these two results.

What the relationship really is, we shall now discuss.

On then to frame 33.

6. The convolution theorem

If $\mathcal{L}^{-1}\{f(s)\} = \mathrm{F}(t)$ and $\mathcal{L}^{-1}\{g(s)\} = \mathrm{G}(t)$, then we wish to investigate $\mathcal{L}^{-1}\{f(s).g(s)\}$.

From the definition, $$f(s) = \int_0^\infty e^{-st}.\mathrm{F}(t)dt$$

and $$g(s) = \int_0^\infty e^{-sT}.\mathrm{G}(T)dT$$

$$\therefore f(s).g(s) = \int_0^\infty e^{-sT}.f(s).\mathrm{G}(T)dT \qquad \ldots \quad \text{(i)}$$

Consider now the product $e^{-sT}.f(s)$ included in the right-hand side expansion.

Now $\mathcal{L}^{-1}\{f(s)\} = \mathrm{F}(t)$

$\therefore \mathcal{L}^{-1}\{e^{-sT}.f(s)\} = \mathrm{F}(t-T).\mathrm{H}(t-T)$ from our work on the unit step function.

$$\therefore e^{-sT}.f(s) = \int_0^\infty e^{-st}\,\mathrm{F}(t-T).\mathrm{H}(t-T)\,dt$$

If we substitute this in (i), we get

$$f(s).g(s) = \int_0^\infty \int_0^\infty e^{-st}\,\mathrm{F}(t-T).\mathrm{H}(t-T).\mathrm{G}(T)\,dt\,dT$$

Since $$\left.\begin{aligned} \mathrm{H}(t-T) &= 0 \quad \text{for} \quad 0 < t < T \\ &= 1 \quad \text{for} \quad t \geqslant T \end{aligned}\right\}$$

$$f(s).g(s) = \int_0^\infty \int_T^\infty e^{-st}\,\mathrm{F}(t-T).\mathrm{G}(T)\,dt\,dT$$

This expression represents integration over the shaded portion of the first quadrant, as shown.

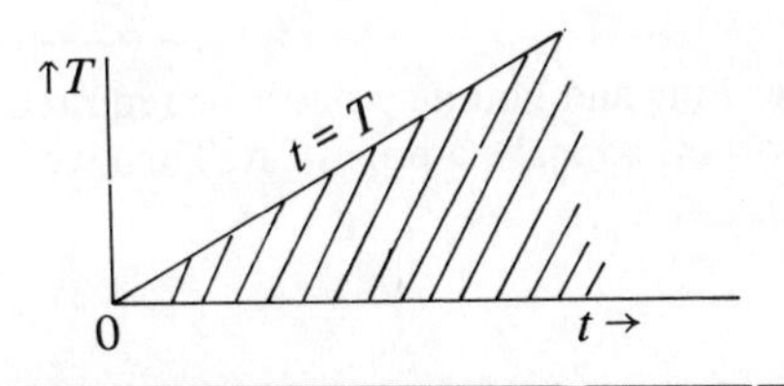

[over

To find this area, we could do one of two things:

(a)

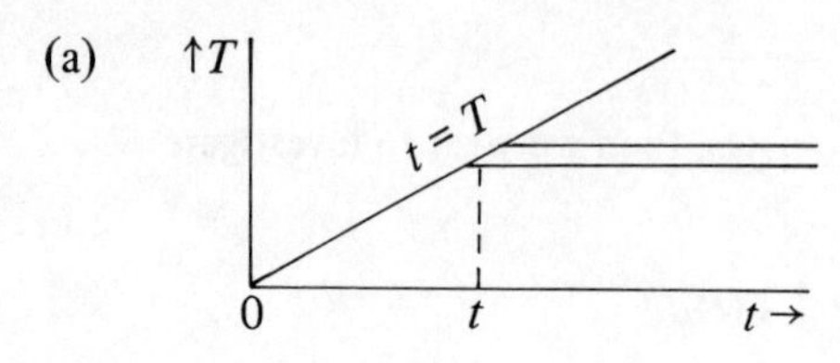

Find the area of a horizontal strip from $t = T$ to $t = \infty$, and then sum all such strips from $T = 0$ to $T = \infty$.

(b)

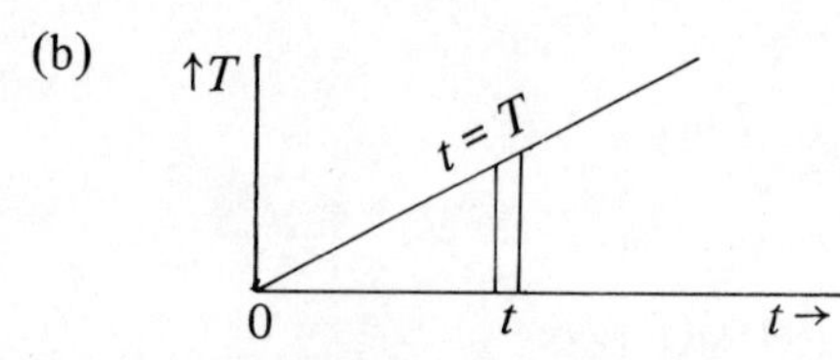

Start with a vertical strip extending from $T = 0$ to $T = t$, and then sum all such strips from $t = 0$ to $t = \infty$.

Starting with the vertical strip, we obtain

$$f(s)\,.\,g(s) = \int_0^\infty \int_0^t e^{-st}\,G(T)\,.\,F(t-T)\,dT\,dt$$

$$= \int_0^\infty e^{-st}\left\{\int_0^t G(T)\,.\,F(t-T)\,dT\right\}dt$$

Now the right-hand side is the Laplace transform of $\int_0^t G(T)\,.\,F(t-T)\,dT$.

$\therefore$ We suddenly arrive at this important result:

If $\mathcal{L}^{-1}\{f(s)\} = F(t)$ and $\mathcal{L}^{-1}\{g(s)\} = G(t)$

then $$\underline{\mathcal{L}^{-1}\{f(s)\,.\,g(s)\} = \int_0^t G(T)\,.\,F(t-T)\,dT} \qquad \ldots \quad \text{VI}$$

or we could equally well have

$$\underline{\mathcal{L}^{-1}\{f(s)\,.\,g(s)\} = \int_0^t F(T)\,.\,G(t-T)\,dT}$$

The proof is rather long and you may never be required to reproduce it in detail, but the result is important so make a note of it. Then we can apply it to one or two examples.

Now on to frame 34.

34

Example 1. To find $\mathcal{L}^{-1}\left\{\frac{1}{s+2}\cdot\frac{s}{s^2+4}\right\}$

Let $\quad f(s) = \frac{s}{s^2+4} \qquad \therefore\ F(t) = \cos 2t$

$\quad g(s) = \frac{1}{s+2} \qquad \therefore\ G(t) = e^{-2t}$

$$\therefore\ \mathcal{L}^{-1}\{f(s)\,.\,g(s)\} = \int_0^t \cos 2T\,.\,e^{-2(t-T)}\,.\,dT$$

$$= \ldots\ldots\ldots\ldots\ldots\ldots\ldots\ldots\ldots\ldots\ldots\ldots\ldots\ldots\ldots$$

35

$$\boxed{\frac{1}{4}\left\{\cos 2t + \sin 2t - e^{-2t}\right\}}$$

Here is the working:

$$\mathcal{L}^{-1}\{f(s).g(s)\} = \int_0^t \cos 2T \,.\, e^{-2(t-T)}\, dT$$

$$= \mathcal{R}\int_0^t e^{j2T} \,.\, e^{-2(t-T)}\, dT$$

$$= \mathcal{R}\, e^{-2t} \,.\, \left\{\int_0^t e^{j2T} \,.\, e^{2T} \,.\, dT\right\}$$

$$= \mathcal{R}\, e^{-2t} \,.\, \int_0^t e^{2(1+j)T}\, dT$$

$$= \mathcal{R}\, e^{-2t} \left[\frac{e^{2(1+j)T}}{2(1+j)}\right]_0^t$$

$$= \mathcal{R}\, e^{-2t} \,.\, \frac{1}{2(1+j)}\left\{e^{2(1+j)t} - 1\right\}.$$

$$= \mathcal{R}\, e^{-2t} \,.\, \frac{1}{2(1+j)}\left\{e^{2t} \,.\, e^{j2t} - 1\right\}$$

$$= \mathcal{R}\, e^{-2t} \,.\, \frac{1}{2(1+j)}\left\{e^{2t}(\cos 2t + j\sin 2t) - 1\right\}$$

$$= \mathcal{R}\, e^{-2t} \,.\, \frac{1-j}{4}\left\{e^{2t}(\cos 2t + j\sin 2t) - 1\right\}$$

$$= \frac{e^{-2t}}{4}\left\{e^{2t}\cos 2t + e^{2t}\sin 2t - 1\right\}$$

$$= \underline{\frac{1}{4}\left\{\cos 2t + \sin 2t - e^{-2t}\right\}}$$

Now another.

Example 2. Find $\mathcal{L}^{-1}\left\{\dfrac{s^2}{(s^2+a^2)^2}\right\}$.

We can think of this as $\mathcal{L}^{-1}\left\{\dfrac{s}{s^2+a^2}\cdot\dfrac{s}{s^2+a^2}\right\}$ and, of course,

$$\mathcal{L}^{-1}\left\{\frac{s}{s^2+a^2}\right\}=\cos at.$$

So the required function of t is:

....

Apply the convolution theorem, as before.

36

$$\frac{1}{2a}\left\{\text{at}\cos at + \sin at\right\}$$

Here are the details:

$$\mathcal{L}^{-1}\left\{\frac{s}{s^2+a^2}\right\} = \cos at$$

$$\mathcal{L}^{-1}\left\{\frac{s}{s^2+a^2}\cdot\frac{s}{s^2+a^2}\right\} = \int_0^t \cos a(t-T)\,.\,\cos \text{a}T\,dT$$

$$= \frac{1}{2}\int_0^t\left\{\cos at + \cos a(t-2T)\right\}dT$$ using $2\cos\text{A}\cos\text{B} = \cos(\text{A}+\text{B}) + \cos(\text{A}-\text{B})$

$$= \frac{1}{2}\left[T\cos at + \frac{\sin a(t-2T)}{-2a}\right]_0^t$$

$$= \frac{1}{2}\left\{\left(t\cos at - \frac{\sin a(-t)}{2a}\right) - \left(0 - \frac{\sin at}{2a}\right)\right\}$$

$$= \frac{1}{2}\left\{t\cos at + \frac{\sin at}{2a} + \frac{\sin at}{2a}\right\}$$

$$= \frac{1}{2a}\left\{at\cos at + \sin at\right\}$$

Now on to Example 3.

37

Example 3. This is a short one but a very important one.

Consider $\mathcal{L}^{-1}\left\{\frac{1}{s}.f(s)\right\}$ Now $\mathcal{L}^{-1}\left\{\frac{1}{s}\right\}=1$. $\mathcal{L}^{-1}\{f(s)\}=F(t)$

$$\mathcal{L}^{-1}\left\{\frac{1}{s}.f(s)\right\} = \int_0^t 1\,.\,F(T)\,dT$$

$$\therefore\ \mathcal{L}^{-1}\left\{\frac{1}{s}.f(s)\right\} = \int_0^t F(T)\,dT \qquad \ldots \quad \text{VII}$$

Note this result. It is a simple example of the convolution theorem, but useful nevertheless.

Example 4. Find $\mathcal{L}^{-1}\left\{\frac{1}{s(s^2-4)}\right\}$.

Complete the work and then check with the next frame.

38

$$\boxed{\frac{1}{4}\left\{\cosh 2t - 1\right\}}$$

for, using the result VII above,

$$\mathcal{L}^{-1}\left\{\frac{1}{s}.\frac{1}{s^2-4}\right\} = \int_0^t 1\,.\,\frac{1}{2}\sinh 2T\,dT = \frac{1}{2}\left[\frac{\cosh 2T}{2}\right]_0^t$$

$$= \frac{1}{4}\left\{\cosh 2t - 1\right\}$$

39 We have dealt with a number of theorems and although we can largely manage without them, they are useful and well worth knowing. Here is a summary of the results we have established.

Summary

1. *Laplace transform of* $\int_0^t \mathrm{F}(t)\,dt$

$$\text{If } y = \int_0^t \mathrm{F}(t)\,dt \quad \text{then} \quad \mathcal{L}\left\{\int_0^t \mathrm{F}(t)\,dt\right\} = \frac{1}{s}\left\{f(s) + y_0\right\}$$

With zero initial conditions $\mathcal{L}\left\{\int_0^t \mathrm{F}(t)\,dt\right\} = \frac{1}{s}.\,f(s).$

2. *Heaviside expansion theorem*

$$f(s) = \frac{\mathrm{P}(s)}{\mathrm{Q}(s)} = \frac{\mathrm{A}_1}{s-a_1} + \frac{\mathrm{A}_2}{s-a_2} + \ldots + \frac{\mathrm{A}_k}{s-a_k} + \ldots + \frac{\mathrm{A}_n}{s-a_n}.$$

$$\mathrm{A}_k = \frac{\mathrm{P}(a_k)}{\mathrm{Q}'(a_k)}$$

$$\mathrm{F}(t) = \frac{\mathrm{P}(a_1)}{\mathrm{Q}'(a_1)}.\,\mathrm{e}^{a_1 t} + \frac{\mathrm{P}(a_2)}{\mathrm{Q}'(a_2)}.\,\mathrm{e}^{a_2 t} + \ldots + \frac{\mathrm{P}(a_k)}{\mathrm{Q}'(a_k)}.\,\mathrm{e}^{a_k t} + \ldots$$

3. *Heaviside series expansion* (for small values of t)

$f(s) = \frac{\mathrm{P}(s)}{\mathrm{Q}(s)}$. Q($s$) of degree n: P(s) of lower degree.

Divide numerator and denominator by s^n and expand in powers of $\frac{1}{s}$.

4. *Initial value theorem*

$$\lim_{t\to 0}\{\mathrm{F}(t)\} = \lim_{s\to\infty}\{s.\,f(s)\}$$

5. *Final value theorem*

$$\lim_{t\to\infty}\{\mathrm{F}(t)\} = \lim_{s\to 0}\{s.\,f(s)\}$$

6. *Convolution theorem*

$$\mathcal{L}^{-1}\{f(s).(gs)\} = \int_0^t \mathrm{F}(T).\mathrm{G}(t-T)\,dT \qquad \text{where} \qquad \begin{cases} f(s) = \mathcal{L}\{\mathrm{F}(t)\} \\ g(s) = \mathcal{L}\{\mathrm{G}(t)\} \end{cases}$$

$$= \int_0^t \mathrm{G}(T).\mathrm{F}(t-T)\,dT$$

Also $$\mathcal{L}^{-1}\left\{\frac{1}{s}.f(s)\right\} = \int_0^t \mathrm{F}(T)\,dT.$$

And finally we come to the test exercise.

40

As usual, the following problems are based directly on the work we have covered in the programme. Do them all: they are quite straightforward.

Text Exercise–VIII

1. If $L\dfrac{di}{dt} + \dfrac{1}{C}\displaystyle\int_0^t i\,dt = F(t)$, find an expression for i in terms of t when $L = 2$, $C = 0{\cdot}1$ and $F(t) = 1 - e^{-2t}$. At $t = 0$, $i = 0$ and the charge on C is zero.

2. Apply the Heaviside expansion theorem to determine $F(t)$, given that
$$f(s) = \frac{P(s)}{Q(s)} = \frac{s+1}{s^3 + s^2 - 20s}.$$

3. If $\mathcal{L}\{F(t)\} = f(s) = \dfrac{s-3}{s^2 - s + 2}$, determine an expression for $F(t)$ for small values of t by using the series expansion method.

4. The current i at time t is given by

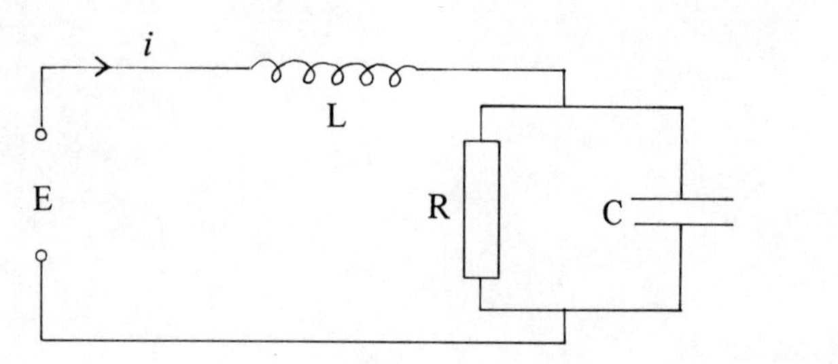

$$\bar{i} = \frac{E}{s\left\{Ls + \dfrac{R}{1 + RCs}\right\}}$$

Determine (i) the value of i as $t \to 0$.
(ii) the value of i as $t \to \infty$.

5. Find $\mathcal{L}^{-1}\left\{\dfrac{4s}{(s^2+4)^2}\right\}$ by writing the transform in the form $\left\{\dfrac{4}{s^2+4}\cdot\dfrac{s}{s^2+4}\right\}$ and applying the convolution theorem.

Further Problems – VIII

1. Apply the Heaviside expansion theorem to determine the inverse transforms of the following:

(a) $\dfrac{1}{s(s+1)(s+2)(s+3)}$

(b) $\dfrac{2s^2-4}{(s+1)(s-2)(s-3)}$

(c) $\dfrac{3s+1}{(s-1)(s^2+1)}$

(d) $\dfrac{s-1}{(s+3)(s^2+2s+2)}$

(e) $\dfrac{s^2-3}{(s+2)(s-3)(s^2+2s+5)}$

2. Use the convolution theorem to determine the inverse transforms of:

(a) $\dfrac{1}{(s+2)^2(s-2)}$

(b) $\dfrac{s}{(s-1)(s^2+1)}$

(c) $\dfrac{2}{s^3(s^2+1)}$

(d) $\dfrac{1}{s^2(s+2)}$

(e) $\dfrac{s}{(s^2+1)^2}$

(f) $\dfrac{1}{s^2(s+1)^2}$

(g) $\dfrac{1}{(s^2+1)^3}$

APPLICATIONS
WORKED EXAMPLES

Mathematical models

Having worked through the foregoing programmes, we can see that the use of Laplace transforms is very largely directed to the solution of differential equations and is especially useful in those cases where the conditions at $t = 0$ are given or can be obtained by knowledge of the problem to which the equation refers.

In dealing with applications from various fields, the first essential is to construct a mathematical model of the problem. This merely means that we have to express the particular details of the problem in mathematical terms and to relate them to form a differential equation. At that stage, the solution of the equation, or set of equations, can of course be determined by the methods of Laplace transforms that we have established in the earlier sections of this work.

The interpretation of the problem into mathematical terms is clearly a vital first step in any problem – and one which often causes some difficulty. In each of the following examples, therefore, we shall dwell on this all-important process, after which the solution by the use of transforms will be relatively straightforward.

Electrical applications – networks

Here we construct the mathematical model by applying

(a) Ohm's law: V = R I
(b) Kirchhoff's laws:
  (i) At any junction in a circuit, the sum of the currents is zero,
  (ii) Around a closed circuit, the sum of the voltage drops is zero.

Voltage drop across basic circuit components:

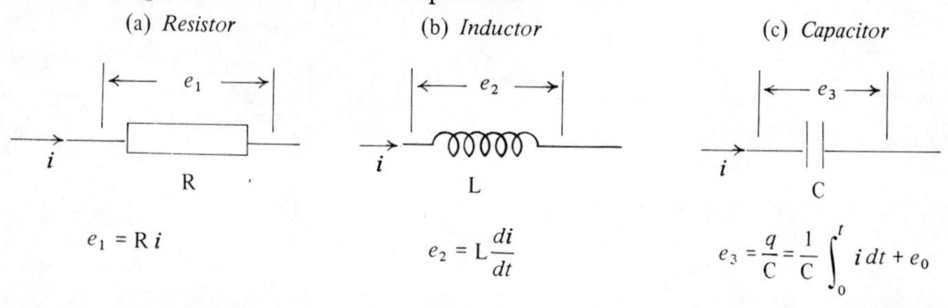

Example 1. Obtain an expression for the current flowing in the circuit shown, given that, at $t = 0$, current and charge are zero.

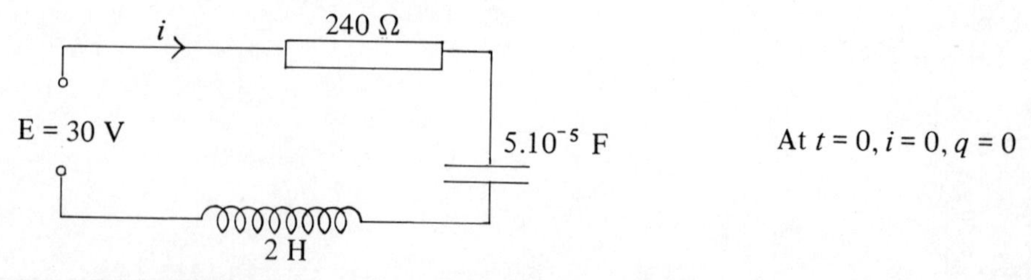

$$240\,i + \frac{1}{5.10^{-5}}\int_0^t i\,dt + 2\frac{di}{dt} = 30$$

Taking transforms:

$$240\,\bar{i} + \frac{1}{5.10^{-5}s}\bar{i} + 2(s\bar{i} - i_0) = \frac{30}{s}. \quad \text{But } i_0 = 0$$

$$\therefore \quad \left\{2s + 240 + \frac{10^5}{5s}\right\}\bar{i} = \frac{30}{s}$$

$$(10s^2 + 1200s + 10^5)\,\bar{i} = 150$$

$$(s^2 + 120s + 10^4)\,i = 15$$

$$\therefore \bar{i} = \frac{15}{s^2 + 120s + 10^4} = \frac{15}{(s+60)^2 + 6400}$$

$$\therefore i = 15\,.\frac{1}{80}\,.\,e^{-60t}\sin 80t \qquad \therefore i = \frac{3}{16}e^{-60t}\sin 80t$$

....

Example 2. For the network shown, find the total current i, given that all initial conditions are zero.

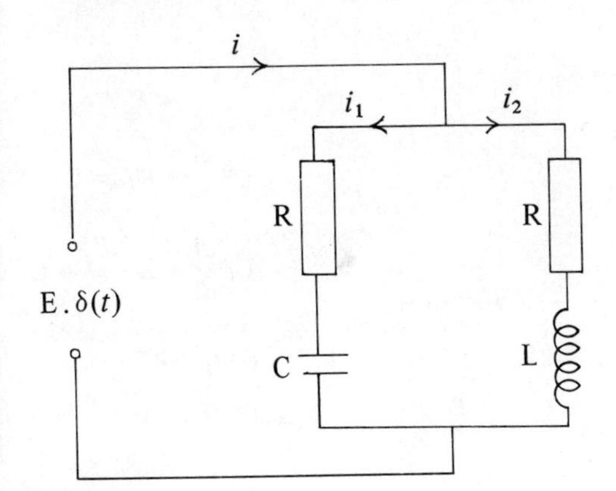

For the L. H. loop:

$$R\,i_1 + \frac{1}{C}\int_0^t i_1\,dt = E\,.\,\delta(t)$$

For the R.H. loop:

$$R\,i_2 + L\frac{di_2}{dt} = E\,.\,\delta(t)$$

All initial conditions are zero.

$$\therefore R\,\bar{i}_1 + \frac{1}{Cs}\bar{i}_1 = E \qquad \therefore \left\{R + \frac{1}{Cs}\right\}\bar{i}_1 = E \qquad \ldots \text{(i)}$$

$$R\bar{i}_2 + L(s\bar{i}_2 - i_{2(0)}) = E \qquad \therefore \quad \{R + Ls\}\,\bar{i}_2 = E \qquad \ldots \quad \text{(ii)}$$

$i = \bar{i}_1 + \bar{i}_2$

From (i) $\bar{i}_1 = \dfrac{E}{R + \dfrac{1}{Cs}} = \dfrac{E}{R}\left\{\dfrac{1}{1 + \dfrac{1}{RCs}}\right\} = \dfrac{E}{R}\left\{\dfrac{s}{s + \dfrac{1}{RC}}\right\}$

From (ii) $\bar{i}_2 = \dfrac{E}{R + Ls} = \dfrac{E}{L}\left\{\dfrac{1}{s + \dfrac{R}{L}}\right\}$

$$\bar{i} = \bar{i}_1 + \bar{i}_2 = \frac{E}{R}\left\{\frac{s}{s + \dfrac{1}{RC}}\right\} + \frac{E}{L}\left\{\frac{1}{s + \dfrac{R}{L}}\right\}$$

$$\bar{i} = \frac{E}{R}\left\{\frac{s + \dfrac{1}{RC}}{s + \dfrac{1}{RC}}\right\} - \frac{E}{R}\cdot\frac{1}{RC}\left\{\frac{1}{s + \dfrac{1}{RC}}\right\} + \frac{E}{L}\left\{\frac{1}{s + \dfrac{R}{L}}\right\}$$

$$\therefore\ i = \frac{E}{R}.\,\delta(t) - \frac{E}{R^2C}e^{-\frac{t}{RC}} + \frac{E}{L}e^{-\frac{Rt}{L}}$$

Example 3. A voltage 200 cos 20t is applied to the network shown. At $t = 0, i = 0$. Find an expression for i in terms of t, for the values given.

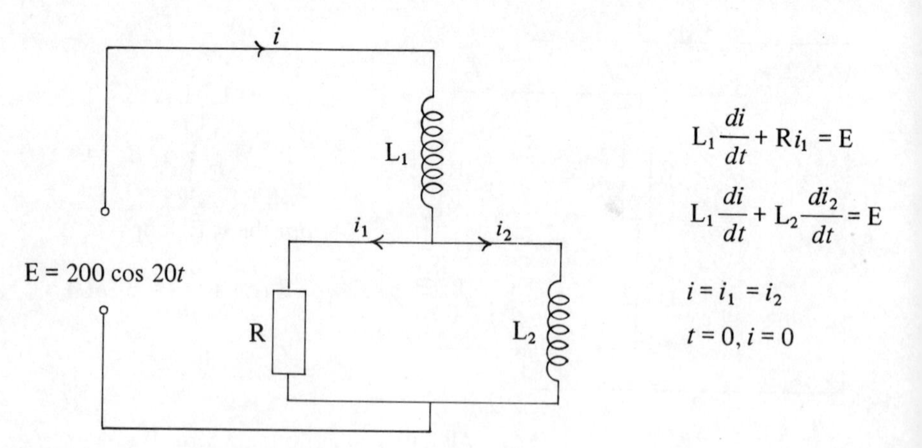

$$L_1\frac{di}{dt} + Ri_1 = E$$

$$L_1\frac{di}{dt} + L_2\frac{di_2}{dt} = E$$

$i = i_1 = i_2$

$t = 0, i = 0$

Consider the case when $L_1 = 2$ H, $L_2 = 4$ H, $R = 80\ \Omega$, $E = 200 \cos 20t$

$$2\frac{di}{dt} + 80\, i_1 = 200 \cos 20t$$

$$2\frac{di}{dt} + 4\frac{di_2}{dt} = 200 \cos 20t$$

$$\begin{cases} 2s\bar{i} + 80\,\bar{i}_1 = 200\dfrac{s}{s^2 + 400} \\[2ex] 2s\bar{i} + 4s\bar{i}_2 = 200\dfrac{s}{s^2 + 400} \qquad \bar{i}_2 = \bar{i} - \bar{i}_1 \end{cases}$$

$$\therefore\ 2s\bar{i} + 4s(\bar{i} - \bar{i}_1) = 200\frac{s}{s^2 + 400}$$

$$\begin{cases} 2s^2\bar{i} + 80s\bar{i}_1 = 200\dfrac{s^2}{s^2 + 400} \\[2ex] 20(2s + 4s)\bar{i} - 80s\bar{i}_1 = 4000\dfrac{s}{s^2 + 400} \end{cases}$$

$$\therefore\ (2s^2 + 120s)\bar{i} = \frac{200s^2 + 4000s}{s^2 + 400}$$

$$\therefore\ \bar{i} = \frac{200s(s + 20)}{s^2 + 400} \cdot \frac{1}{2s(s + 60)} = \frac{100(s + 20)}{(s + 60)(s^2 + 400)}$$

$$= 100 \cdot \frac{(s + 60) - 40}{(s + 60)(s^2 + 400)}$$

$$= 100\left\{\frac{1}{s^2 + 400} - 40 \cdot \frac{1}{(s + 60)(s^2 + 400)}\right\}$$

$$= 100\left\{\frac{1}{s^2 + 400} - 40\left[\frac{1/4000}{s + 60} + \frac{-s/4000 + 3/200}{s^2 + 400}\right]\right\}$$

$$= 40\frac{1}{s^2 + 400} - \frac{1}{s + 60} + \frac{s}{s^2 + 400}$$

$$\therefore\ \underline{i = 2 \sin 20t + \cos 20t - e^{-60t}}$$

Example 4. In the coupled circuits shown below, $R_1 = 5\ \Omega$; $R_2 = 20\ \Omega$; $L_1 = 2$ H; $L_2 = 3$H; M = 1H; E = 100 V. At $t = 0$, $i_1 = i_2 = 0$. Find the secondary current, i_2, at time t.

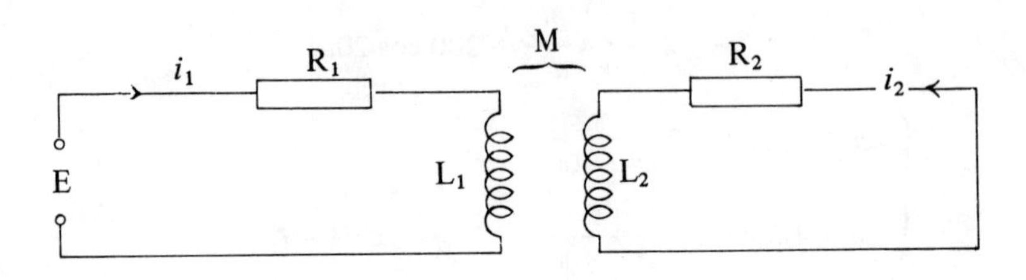

Primary: $$L_1 \frac{di_1}{dt} + R_1 i_1 + M \frac{di_2}{dt} = E$$

Secondary: $$L_2 \frac{di_2}{dt} + R_2 i_2 + M \frac{di_1}{dt} = 0$$

$$\therefore \begin{cases} 2s\bar{i}_1 + 5\bar{i}_1 + 1 . s\bar{i}_2 = \dfrac{100}{s} \\ 3s\bar{i}_2 + 20\bar{i}_2 + 1 . s\bar{i}_1 = 0 \end{cases}$$

$$\therefore \begin{cases} (2s+5)\bar{i}_1 + s\bar{i}_2 = \dfrac{100}{s} \\ s\bar{i}_1 + (3s+20)\bar{i}_2 = 0 \end{cases}$$

$$\therefore \begin{cases} s(2s+5)\bar{i}_1 + s^2\bar{i}_2 = 100 \\ s(2s+5)\bar{i}_1 + (2s+5)(3s+20)\bar{i}_2 = 0 \end{cases}$$

$$\therefore\ (5s^2 + 55s + 100)\bar{i}_2 = -100$$

$$(s^2 + 11s + 20)\bar{i}_2 = -20$$

$$\left\{\left(s + \frac{11}{2}\right)^2 - \frac{41}{4}\right\} \bar{i}_2 = -20$$

$$\therefore\ \bar{i}_2 = -20 \,.\, \frac{1}{\left\{s+\frac{11}{2}-\frac{\sqrt{41}}{2}\right\}\left\{s+\frac{11}{2}+\frac{\sqrt{41}}{2}\right\}}$$

$$= -20\left[\frac{1}{\sqrt{41}}\cdot\frac{1}{s+\frac{11}{2}-\frac{\sqrt{41}}{2}} - \frac{1}{\sqrt{41}}\cdot\frac{1}{s+\frac{11}{2}+\frac{\sqrt{41}}{2}}\right]$$

$$= \frac{20}{\sqrt{41}}\left\{\frac{1}{s+\frac{11+\sqrt{41}}{2}} - \frac{1}{s+\frac{11-\sqrt{41}}{2}}\right\}$$

$$i_2 = \frac{20}{\sqrt{41}}\left\{e^{-\frac{1}{2}(11+\sqrt{41})t} - e^{-\frac{1}{2}(11-\sqrt{41})t}\right\}$$

Mechanical Systems

The main units in forming a mathematical model of a mechanical problem are as listed below.

(a) *Spring*

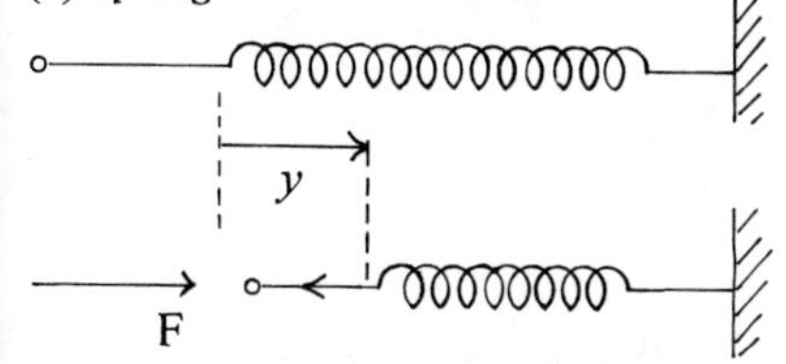

If a spring is compressed (or extended) a distance y by an applied force F, then

$$F = ky$$

where k is the spring constant.

(b) *Damper*

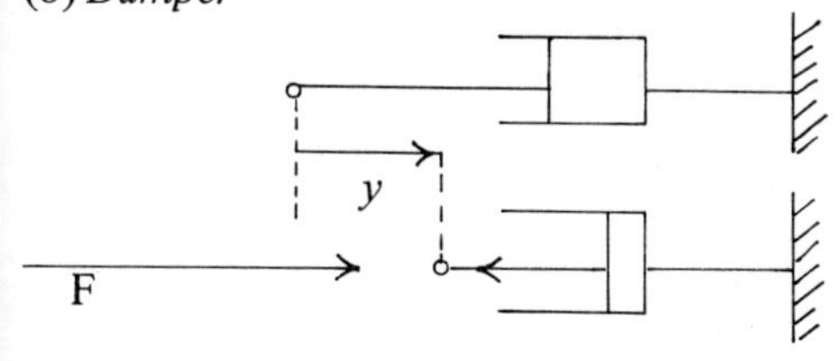

The resistive force exerted by the damper is proportional to the velocity

i.e. $\quad F = B\dot{y}$

where B is the coefficient of viscous damping.

Motion in a resistive medium is also affected by a retarding force proportional to the velocity and acts against the direction of motion.

(c) *Acceleration*

The sum of the forces acting on a mass is proportional to the acceleration of the mass,

i.e. $$F = M\ddot{y}$$

Example 1. A mass M is attached to a spring in equilibrium and rests on a frictionless plane. If the mass is displaced through a distance c and released, determine the resulting equation of motion and hence obtain an expression for the displacement at time t

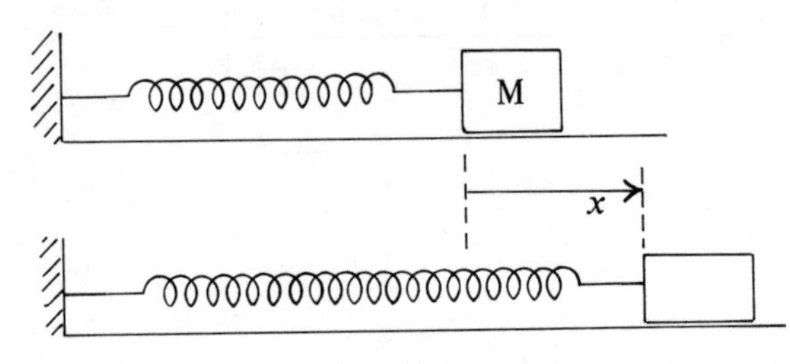

At time t, the force on the mass is the tension in the spring, i.e. kx.
The acceleration of the mass = $\ddot{x}$

$$M\ddot{x} = -kx \qquad \text{and, at} \qquad t = 0, x = c \qquad \text{and} \qquad \dot{x} = 0.$$

$$\therefore\ \ddot{x} + \frac{k}{M}x = 0 \qquad \therefore\ (s^2\bar{x} - sx_0 - x_1) + \frac{k}{M}\bar{x} = 0$$

$$\therefore\ \left(s^2 + \frac{k}{M}\right)\bar{x} - cs = 0 \qquad \therefore\ \bar{x} = \frac{cs}{s^2 + \frac{k}{M}}$$

$$\therefore\ x = c\cos\sqrt{\frac{k}{M}}\,.\,t$$

$\therefore$ The mass performs simple harmonic motion, with amplitude c and frequency

$$f = \frac{1}{2\pi}\sqrt{\frac{k}{M}} \quad \text{vibrations/sec.}$$

Example 2. A vertical coiled spring is extended 5 cm when a mass of 2 kg is attached to it. From its equilibrium position, the mass is pulled down a distance of 4 cm and released. If the damping factor is 2 and is proportional to the instantaneous value of the velocity, determine the expression for the displacement at time t.

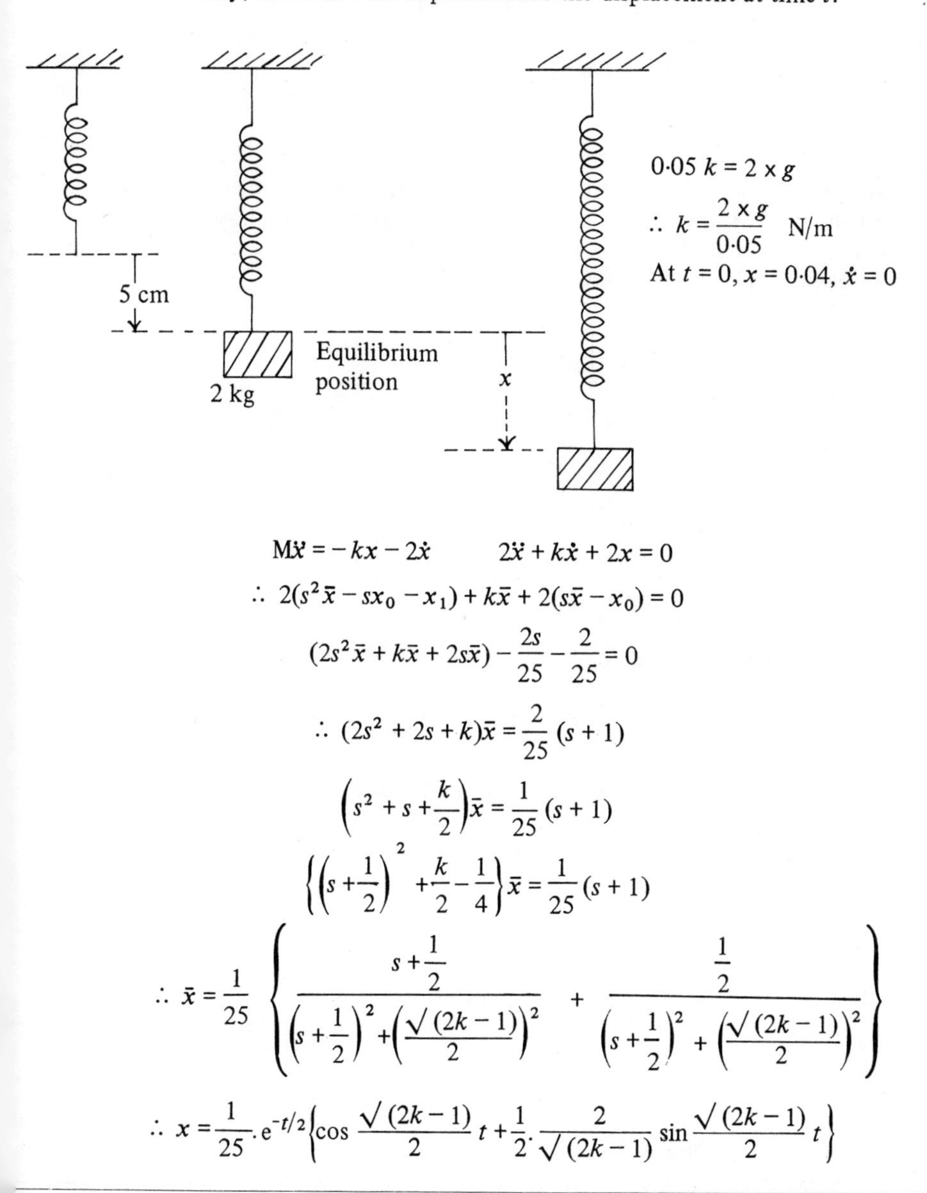

$$0{\cdot}05\,k = 2 \times g$$

$$\therefore\; k = \frac{2 \times g}{0{\cdot}05}\ \text{N/m}$$

At $t = 0$, $x = 0{\cdot}04$, $\dot{x} = 0$

$$M\ddot{x} = -kx - 2\dot{x} \qquad 2\ddot{x} + k\dot{x} + 2x = 0$$

$$\therefore\; 2(s^2\bar{x} - sx_0 - x_1) + k\bar{x} + 2(s\bar{x} - x_0) = 0$$

$$(2s^2\bar{x} + k\bar{x} + 2s\bar{x}) - \frac{2s}{25} - \frac{2}{25} = 0$$

$$\therefore\; (2s^2 + 2s + k)\bar{x} = \frac{2}{25}(s+1)$$

$$\left(s^2 + s + \frac{k}{2}\right)\bar{x} = \frac{1}{25}(s+1)$$

$$\left\{\left(s + \frac{1}{2}\right)^2 + \frac{k}{2} - \frac{1}{4}\right\}\bar{x} = \frac{1}{25}(s+1)$$

$$\therefore\; \bar{x} = \frac{1}{25}\left\{\frac{s + \frac{1}{2}}{\left(s + \frac{1}{2}\right)^2 + \left(\frac{\sqrt{(2k-1)}}{2}\right)^2} + \frac{\frac{1}{2}}{\left(s + \frac{1}{2}\right)^2 + \left(\frac{\sqrt{(2k-1)}}{2}\right)^2}\right\}$$

$$\therefore\; x = \frac{1}{25}\cdot e^{-t/2}\left\{\cos\frac{\sqrt{(2k-1)}}{2}\,t + \frac{1}{2}\cdot\frac{2}{\sqrt{(2k-1)}}\sin\frac{\sqrt{(2k-1)}}{2}\,t\right\}$$

Now $\quad k = \dfrac{2 \times 9{\cdot}81}{0{\cdot}05}$ N/m = 392·4 $\quad \therefore\ 2k - 1 = 783{\cdot}8 \quad \therefore\ \sqrt{(2k-1)} \doteq 28$

$$\therefore\ x = \frac{1}{25}\,e^{-t/2}\left\{\cos 14t + \frac{1}{28}\sin 14t\right\} \quad \ldots\ldots\ldots\ldots\ x \text{ in metres}$$

$$x = 4\,e^{-t/2}\left\{\cos 14t + \frac{1}{28}\sin 14t\right\} \quad \ldots\ldots\ldots\ldots\ x \text{ in cm.}$$

Example 3. A mass M_1 is supported by a spring-damper system and itself supports a second mass M_2 on a spring as shown in the diagram. A unit impulse is applied to M_1 at $t = 0$ when $x = y = 0$. From the differential equations expressed in Laplace transforms, to determine x and y in terms of t.

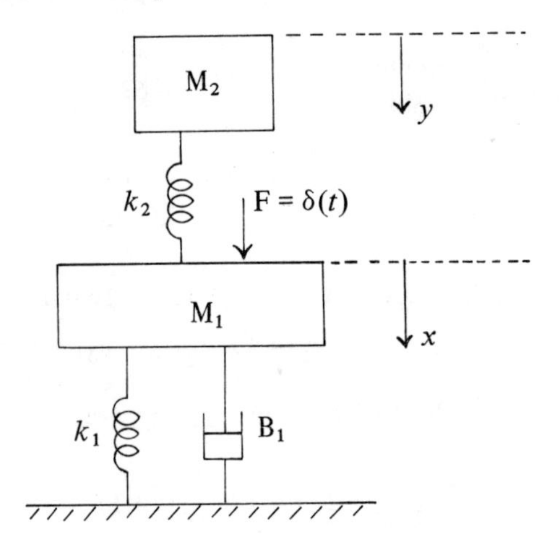

(a) Forces on M_1: $\quad F + k_2(y - x) - k_1 x - B_1\dot{x}$

$$\therefore\ M_1\ddot{x} = \delta(t) + k_2(y - x) - k_1 x - B_1\dot{x}$$

(b) Forces on M_2: $\quad k_2(x - y)$

$$\therefore\ M_2\ddot{y} = k_2(x - y)$$

$$\left.\begin{aligned} M_1\ddot{x} &= \delta(t) + k_2(y - x) - k_1 x - B_1\dot{x} \\ M_2\ddot{y} &= k_2(x - y) \end{aligned}\right\} \quad \text{Zero initial conditions.}$$

$$\left.\begin{aligned} M_1 s^2\bar{x} &= 1 + k_2\bar{y} - k_2\bar{x} - k_1\bar{x} - B_1 s\bar{x} \\ M_2 s^2\bar{y} &= k_2\bar{x} - k_2\bar{y} \end{aligned}\right\}$$

$$\left.\begin{aligned}(M_1s^2 + B_1s + k_1 + k_2)\bar{x} - k_2\bar{y} &= 1\\ k_2\bar{x} - (M_2s^2 + k_2)\bar{y} &= 0\end{aligned}\right\}$$

These equations can then be solved by elimination in the usual way.

Structures – loaded beams

If a horizontal beam of length $2c$ is subjected to a vertical load function, $F(x)$, the resulting deflection y at a point in the beam, distance x from one end, is given by

$$EI\frac{d^4y}{dx^4} = F(x) \qquad \text{for} \qquad 0 < x < 2c$$

E = modulus of elasticity of material of the beam
I = second moment of area of cross-section of the beam.

At any point along the beam, the bending moment $= EIy''(x)$

and the shearing force $= EIy'''(x)$

Also: (a) for cantilever beam $y = 0$ and $y' = 0$ at point of support.
(b) for simply supported beam $y = 0$ and $y'' = 0$ at point of support.
(c) for free end of beam $y'' = 0$ and $y''' = 0$.

Note: The independent variable is now x instead of t as in previous examples.

Example 1. A beam of length $2c$ is simply supported at each end and carries a uniformly distributed load of W (newtons/metre). Find an expression for the deflection y in terms of x, the distance from one end.

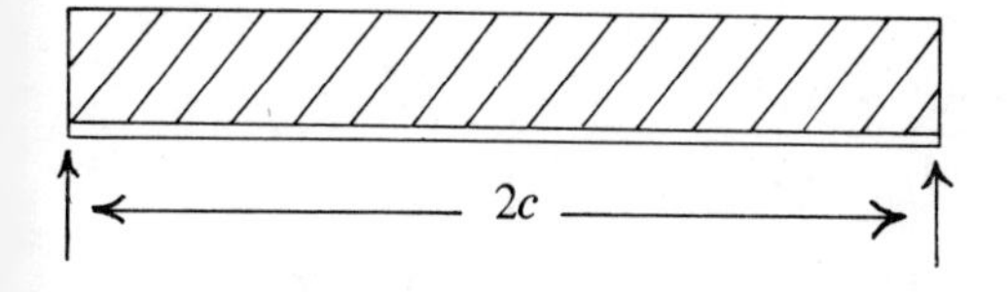

$$EI\frac{d^4y}{dx^4} = W \qquad 0 < x < 2c$$

At $x = 0$, $y = 0$, $y'' = 0$

At $x = 2c$, $y = 0$, $y'' = 0$

$$\therefore\ s^4\bar{y} - s^3y_0 - s^2y_1 - sy_2 - y_3 = \frac{W}{EIs}$$

$$y_0 = 0; \quad y_2 = 0 \qquad \therefore\ s^4\bar{y} - s^2y_1 - y_3 = \frac{W}{EIs}$$

Let $y_1 = A$; $y_3 = B$ $\qquad \therefore\ s^4 y = As^2 + B + \dfrac{W}{EIs}$

$$\therefore\ \bar{y} = \frac{A}{s^2} + \frac{B}{s^4} + \frac{W}{EI}\cdot\frac{1}{s^5}$$

$$\therefore\ y = Ax + \frac{Bx^3}{6} + \frac{W}{EI}\cdot\frac{x^4}{24} \qquad \ldots \text{(i)}$$

At $x = 2c$, $\quad y = 0$

$$\therefore\ 0 = 2Ac + \frac{B8c^3}{6} + \frac{W}{EI}\cdot\frac{16c^4}{24}$$

$$0 = 2Ac + \frac{4Bc^3}{3} + \frac{W}{EI}\cdot\frac{2c^4}{3} \qquad \ldots \text{(ii)}$$

At $x = 2c$, $\quad y'' = 0$

$$y' = A + \frac{Bx^2}{2} + \frac{W}{EI}\cdot\frac{x^3}{6}$$

$$y'' = Bx + \frac{W}{EI}\,\frac{x^2}{2}$$

$$\therefore\ 2Bc + \frac{W}{EI}\cdot\frac{4c^2}{2} = 0 \qquad \therefore\ B = -\frac{Wc}{EI}$$

Substitute in (ii)

$$0 = 2Ac - \frac{4Wc^4}{3EI} + \frac{2Wc^4}{3EI} \qquad \therefore\ A = \frac{Wc^3}{3EI}$$

$\therefore$ (i) becomes

$$y = \frac{Wc^3x}{3EI} - \frac{Wcx^3}{6EI} + \frac{Wx^4}{24EI}$$

$$\therefore\ \underline{y = \frac{Wx}{24EI}\left\{8c^3 - 4cx^2 + x^3\right\}} \qquad 0 < x < 2c$$

Example 2. A uniform light horizontal beam PQ, of length $2c$ and supported at P, carries a load which decreases uniformly from w (newtons/metre) at P to zero at $x = c$. A point load F (newtons) occurs at a distance b from P. Find an expression for the displacement at any point in the beam in terms of x.

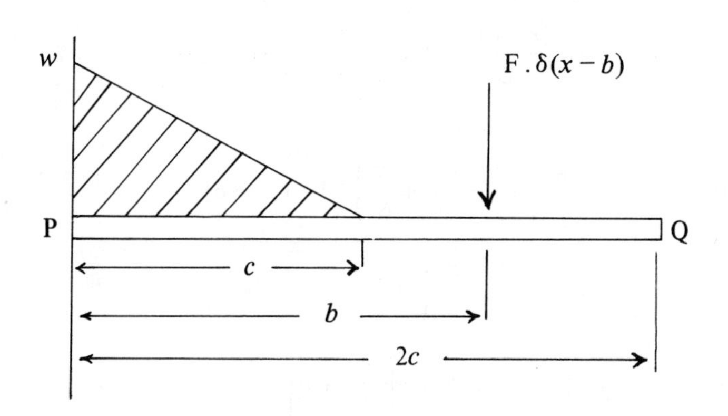

$$\mathrm{EI}\frac{d^4y}{dx^4} = w - \frac{w}{c}x - \left(w - \frac{w}{c}x\right).\mathrm{H}(x-c) + \mathrm{F}.\delta(x-b)$$

$$= \frac{w}{c}\left\{c - x - (c-x).\mathrm{H}(x-c)\right\} + \mathrm{F}.\delta(x-b)$$

$$= \frac{w}{c}\left\{c - x + (x-c).\mathrm{H}(x-c)\right\} + \mathrm{F}.\delta(x-b)$$

Cantilever: $\therefore$ At $x = 0$, $y = 0$, $y' = 0$

At $x = 2c$, $y'' = 0$, $y''' = 0$.

$$\mathrm{EI}\left\{s^4\bar{y} - s^3y_0 - s^2y_1 - sy_2 - y_3\right\} = \frac{w}{c}\left\{\frac{c}{s} - \frac{1}{s^2} + \frac{e^{-cs}}{s^2}\right\} + \mathrm{F}.e^{-bs}$$

Let $y_2 = \mathrm{A}$ and $y_3 = \mathrm{B}$

$$\therefore \mathrm{EI}\left\{s^4\bar{y} - \mathrm{A}s - \mathrm{B}\right\} = \frac{w}{c}\left\{\frac{c}{s} - \frac{1}{s^2} + \frac{e^{-cs}}{s^2}\right\} + \mathrm{F}.e^{-bs}$$

$$\therefore \mathrm{EI}\bar{y} = \frac{\mathrm{A}}{s^3} + \frac{\mathrm{B}}{s^4} + \frac{w}{c}\left\{\frac{c}{s^5} - \frac{1}{s^6} + \frac{e^{-cs}}{s^6}\right\} + \mathrm{F}.\frac{e^{-bs}}{s^4}$$

$$\therefore \mathrm{EI}y = \frac{\mathrm{A}x^2}{2} + \frac{\mathrm{B}x^3}{6} + \frac{w}{c}\left\{\frac{cx^4}{24} - \frac{x^5}{120} + \frac{(x-c)^5}{120}.\mathrm{H}(x-c)\right\}$$

$$+ \mathrm{F}.\frac{(x-b)^3}{6}.\mathrm{H}(x-b).$$

$$EIy' = Ax + \frac{Bx^2}{2} + \frac{w}{c}\left\{\frac{cx^3}{6} - \frac{x^4}{24} + \frac{(x-c)^4}{24} \,.\, H(x-c)\right\} + F \,.\, \frac{(x-b)^2}{2} \,.\, H(x-b)$$

$$EIy'' = A + Bx + \frac{w}{c}\left\{\frac{cx^2}{2} - \frac{x^3}{6} + \frac{(x-c)^3}{6} \,.\, H(x-c)\right\} + F(x-b) \,.\, H(x-b)$$

$$EIy''' = B + \frac{w}{c}\left\{cx - \frac{x^2}{2} + \frac{(x-c)^2}{2} \,.\, H(x-c)\right\} + F \,.\, H(x-b)$$

At $x = 2c$, $y'' = 0$ and $y''' = 0$

$$\therefore\ 0 = A + 2Bc + \frac{w}{c}\left\{2c^3 - \frac{4c^3}{3} + \frac{c^3}{6}\right\} + F \,.\, (2c-b)$$

and $$0 = B + \frac{w}{c}\left\{2c^2 - \frac{4c^2}{2} + \frac{c^2}{2}\right\} + F.$$

$$\therefore\ B = -\frac{w}{c} \cdot \frac{c^2}{2} - F \qquad \therefore\ B = -\left[\frac{wc}{2} + F\right]$$

$$0 = A - 2c\left[\frac{wc}{2} + F\right] + \frac{w}{c}\left\{2 - \frac{4}{3} + \frac{1}{6}\right\} c^3 + F(2c-b)$$

$$= A - wc^2 - 2cF + wc^2 \,.\, \frac{5}{6} + F(2c-b).$$

$$= A - \frac{wc^2}{6} - Fb \qquad \therefore\ A = \frac{wc^2}{6} + Fb$$

Substituting the values of A and B into the previous result, we have

$$EIy = \left[\frac{wc^2}{6} + Fb\right]\frac{x^2}{2} - \left[\frac{wc}{2} + F\right]\frac{x^3}{6} + \frac{w}{c}\left[\frac{cx^4}{24} - \frac{x^5}{120} + \frac{(x-c)^5}{210} \,.\, H(x-c)\right] + F \,.\, \frac{(x-b)^3}{6} \,.\, H(x-b).$$

APPENDIX

1. *To prove that* $\lim_{t\to\infty} \{t^n e^{-st}\} = 0. \qquad (s > 0)$

$$e^{st} = 1 + st + \frac{s^2 t^2}{2!} + \frac{s^3 t^3}{3!} + \ldots$$

If $s > 0$ and $t > 0$, then $e^{st} >$ any one term

i.e. $$e^{st} > \frac{s^{n+1} t^{n+1}}{(n+1)!}$$

Dividing both sides by t^n

$$\frac{e^{st}}{t^n} > \frac{s^{n+1}\, t}{(n+1)!}$$

As $t \to \infty$, R.H.S. $\to \infty$. $\quad \therefore \frac{e^{st}}{t^n} \to \infty$

$$\therefore \frac{t^n}{e^{st}} \to 0 \qquad \therefore t^n e^{-st} \to 0$$

$$\lim_{t\to\infty} \{t^n e^{-st}\} = 0 \qquad \ldots \text{I}$$

2. *Theorem 1 – First shift theorem*

$$\mathcal{L}\{F(t)\} = \int_0^\infty e^{-st} F(t)\, dt = f(s)$$

$$\mathcal{L}\{e^{-at} F(t)\} = \int_0^\infty e^{-st} e^{-at} F(t)\, dt$$

$$= \int_0^\infty e^{-(s+a)t} . F(t)\, dt$$

This is the same as $f(s)$ above, with s replaced by $(s + a)$

If $\mathcal{L}\{F(t)\} = f(s)$, then $\mathcal{L}\{e^{at} F(t)\} = f(s + a)$ $\qquad \ldots$ II

3. *Theorem 2 – Multiplying by t^n*

$$\mathcal{L}\{F(t)\} = \int_0^\infty e^{-st} F(t)\,dt = f(s)$$

$$\therefore \frac{d}{ds}\left\{f(s)\right\} = \frac{d}{ds}\int_0^\infty e^{-st} F(t)\,dt$$

$$= \int_0^\infty \frac{\partial}{\partial s} . e^{-st} F(t)\,dt$$

$$= \int_0^\infty -t\, e^{-st} F(t)\,dt$$

$$= -\int_0^\infty e^{-st}\, t\, F(t)\,dt$$

$$= -\mathcal{L}\{t\,F(t)\}$$

$$\underline{\mathcal{L}\{t\,F(t)\} = -\frac{d}{ds}\{f(s)\}}$$

Similarly

$$\mathcal{L}\{t^2\,F(t)\} = \mathcal{L}\{t[t\,F(t)]\}$$

$$= -\frac{d}{ds}\left[-\frac{d}{ds}\{f(s)\}\right]$$

$$\underline{\mathcal{L}\{t^2\,F(t)\} = \frac{d^2}{ds^2}\{f(s)\}} \qquad \text{etc.}$$

Thus, in general:

$$\underline{\mathcal{L}\{t^n\,F(t)\} = (-1)^n . \frac{d^n}{ds^n}\{f(s)\}} \qquad \ldots \text{III}$$

4. *Theorem 3 – Dividing by t*

$$\int_s^\infty f(s)\,ds = \int_s^\infty \int_0^\infty e^{-st}\,F(t)\,dt\,ds$$

$$= \int_0^\infty \int_s^\infty e^{-st}\,F(t)\,ds\,dt$$

$$= \int_0^\infty \left[\frac{e^{-st}\,F(t)}{-t}\right]_s^\infty dt$$

$$= \int_0^\infty \left[\{0\} - \left\{\frac{e^{-st}\,F(t)}{-t}\right\}\right] dt$$

$$= \int_0^\infty e^{-st}\,\frac{F(t)}{t}\,dt = \mathcal{L}\left\{\frac{F(t)}{t}\right\}$$

provided that $\lim_{t\to 0}\left\{\frac{F(t)}{t}\right\}$ exists.

$\therefore$ If $\mathcal{L}\{F(t)\} = f(s)$, then $\mathcal{L}\left\{\frac{F(t)}{t}\right\} = \int_s^\infty f(s)\,ds$ ... IV

Table of Laplace Transforms

$F(t)$	$\mathcal{L}\{F(t)\} = f(s)$
a	$\frac{a}{s}$
e^{at}	$\frac{1}{s-a}$
t	$\frac{1}{s^2}$
$t^n \ (n = 1, 2, 3, \ldots)$	$\frac{n!}{s^{n+1}}$
$\sin at$	$\frac{a}{s^2 + a^2}$
$\cos at$	$\frac{s}{s^2 + a^2}$
$\sinh at$	$\frac{a}{s^2 - a^2}$
$\cosh at$	$\frac{s}{s^2 - a^2}$
$t \sin at$	$\frac{2as}{(s^2 + a^2)^2}$
$t \cos at$	$\frac{s^2 - a^2}{(s^2 + a^2)^2}$
$t \sinh at$	$\frac{2as}{(s^2 - a^2)^2}$
$t \cosh at$	$\frac{s^2 + a^2}{(s^2 - a^2)^2}$
$e^{-at} F(t)$	$f(s + a)$
$t^n F(t)$	$(-1)^n \cdot \frac{d^n}{ds^n} \{f(s)\}$
$\frac{F(t)}{t}$	$\int_s^{\infty} f(s)\, ds$, if $\lim_{t \to 0} \left\{\frac{F(t)}{t}\right\}$ exists
$\frac{\sin at}{t}$	$\tan^{-1} \left\{\frac{a}{s}\right\}$
$H(t - a)$	$\frac{e^{-as}}{s}$
$H(t)$	$\frac{1}{s}$
$\delta(t - a)$	e^{-as}
$\delta(t)$	1
$\dot{y}$	$s\bar{y} - y_0$
$\ddot{y}$	$s^2\bar{y} - sy_0 - y_1$
$\dddot{y}$	$s^3\bar{y} - s^2 y_0 - sy_1 - y_2$

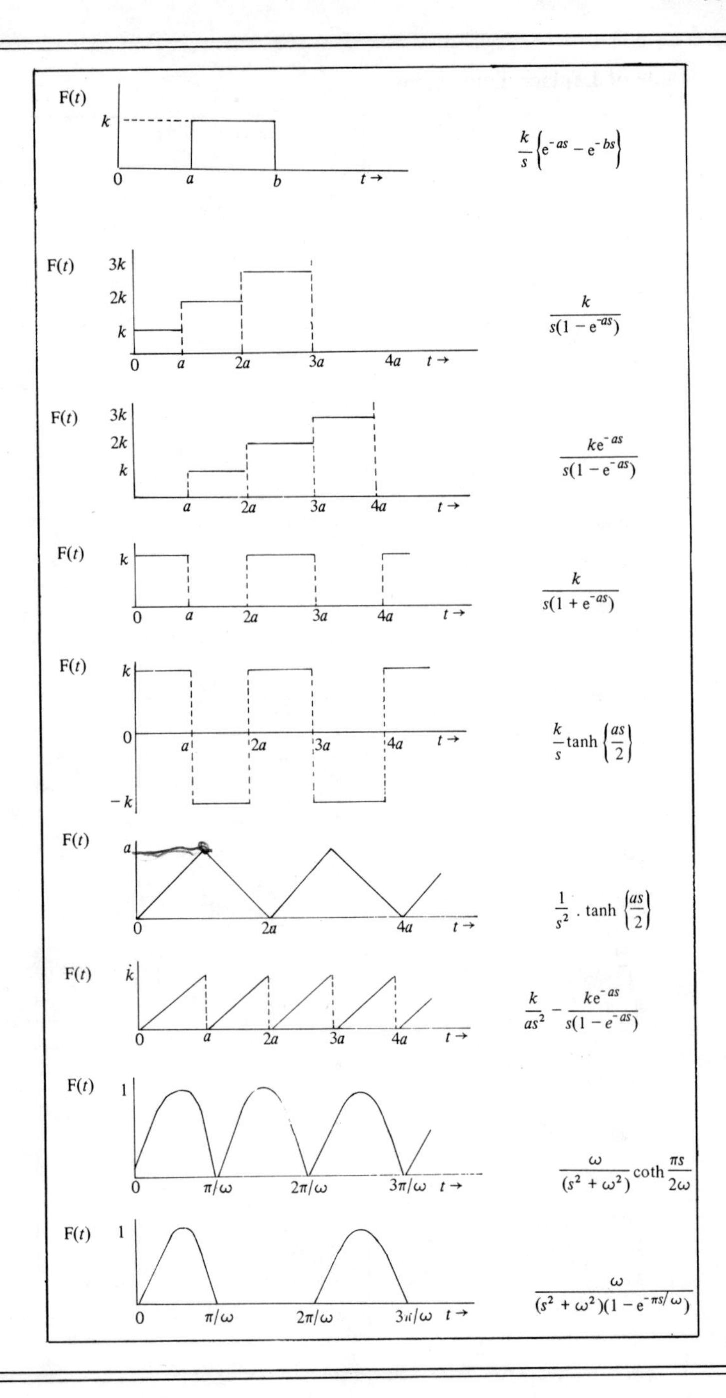
F(t)
k
0
a
b
t →
$\frac{k}{s}\left\{e^{-as} - e^{-bs}\right\}$
F(t)
3k
2k
k
0
a
2a
3a
4a
t →
$\frac{k}{s(1 - e^{-as})}$
F(t)
3k
2k
k
a
2a
3a
4a
t →
$\frac{ke^{-as}}{s(1 - e^{-as})}$
F(t)
k
0
a
2a
3a
4a
t →
$\frac{k}{s(1 + e^{-as})}$
F(t)
k
0
a
2a
3a
4a
t →
−k
$\frac{k}{s}\tanh\left\{\frac{as}{2}\right\}$
F(t)
a
0
2a
4a
t →
$\frac{1}{s^2} \cdot \tanh\left\{\frac{as}{2}\right\}$
F(t)
k
0
a
2a
3a
4a
t →
$\frac{k}{as^2} - \frac{ke^{-as}}{s(1 - e^{-as})}$
F(t)
1
0
π/ω
2π/ω
3π/ω
t →
$\frac{\omega}{(s^2 + \omega^2)} \coth \frac{\pi s}{2\omega}$
F(t)
1
0
π/ω
2π/ω
3π/ω
t →
$\frac{\omega}{(s^2 + \omega^2)(1 - e^{-\pi s/\omega})}$

Table of Inverse Transforms

$f(s)$	$F(t) = \mathcal{L}^{-1}\{f(s)\}$
$\frac{a}{s}$	a
$\frac{1}{s^n}$	$\frac{t^{n+1}}{(n-1)!}$
$\frac{1}{s-a}$	e^{at}
$\frac{1}{(s-a)^n}$	$\frac{t^{n-1}e^{at}}{(n-1)!}$
$\frac{1}{(s-a)(s-b)}$	$\frac{1}{a-b}\{e^{at}-e^{bt}\}$
$\frac{s}{(s-a)(s-b)}$	$\frac{1}{a-b}\{ae^{at}-be^{bt}\}$
$\frac{1}{s^2+a^2}$	$\frac{1}{a}\sin at$
$\frac{s}{s^2+a^2}$	$\cos at$
$\frac{1}{s^2-a^2}$	$\frac{1}{a}\sinh at$
$\frac{s}{s^2-a^2}$	$\cosh at$
$\frac{1}{(s-a)^2+b^2}$	$\frac{1}{b}e^{at}\sin bt$
$\frac{s-a}{(s-a)^2+b^2}$	$e^{at}\cos bt$
$\frac{1}{s(s^2+a^2)}$	$\frac{1}{a^2}\{1-\cos at\}$
$\frac{1}{s^2(s^2+a^2)}$	$\frac{1}{a^3}\{at-\sin at\}$
$\frac{1}{(s^2+a^2)^2}$	$\frac{1}{2a^3}\{\sin at - at\cos at\}$
$\frac{s}{(s^2+a^2)^2}$	$\frac{t}{2a}\sin at$
$\frac{s^2}{(s^2+a^2)^2}$	$\frac{1}{2a}\{\sin at + at\cos at\}$
$\frac{s}{(s^2+a^2)(s^2+b^2)}$ $[a^2 \neq b^2]$	$\frac{1}{b^2-a^2}\{\cos at - \cos bt\}$

ANSWERS

Test Exercise I (Page 30)

1. $\dfrac{25}{s}$
2. $\dfrac{4}{s-3}$
3. $\dfrac{5}{s^2+25}$
4. $\dfrac{2s}{s^2+9}$
5. $\dfrac{s}{s^2-4}$
6. $\dfrac{3}{s^2-1}$
7. $\dfrac{s+6}{s^2+36}$
8. $\dfrac{-(s+20)}{s^2+16}$
9. $\dfrac{24}{s^5}+\dfrac{6}{s^3}-\dfrac{2}{s}$
10. $\dfrac{3}{s^2+4s+13}$
11. $\dfrac{6}{(s-1)^4}$
12. $\dfrac{2(s-2)}{(s-3)^3}$
13. $\dfrac{s+7}{s^2+10s+21}$
14. $\dfrac{4s}{(s^2+4)^2}$
15. $\dfrac{16+12s^2}{(s^2+4)^3}$
16. $\dfrac{1}{2}\ln\left\{\dfrac{s^2+4}{s^2}\right\}$
17. $\dfrac{s-4}{s^2-8s+20}$
18. $\dfrac{s+3}{s^2+6s}$
19. $\dfrac{s^2+4}{(s^2-4)^2}$
20. $\dfrac{1}{2}\ln\left\{\dfrac{s+1}{s-1}\right\}$

Further Problems I (Page 31)

1. $\dfrac{s+2}{s^2+4s+20}$
2. $\dfrac{30s^2+250}{(s^2+25)^3}$
3. $\dfrac{5}{s^2-6s+34}$
4. $\dfrac{4(3s^2-4)}{(s^2+4)^3}$
5. $\dfrac{24}{(s+24)^5}$
6. $(s\cos\theta - w\sin\theta)/(s^2+w^2)$
7. $\dfrac{s^2+2k^2}{s(s^2+4k^2)}$
8. $\dfrac{2n(3s^2-n^2)}{(s^2+n^2)^3}$
9. $(w\cos\theta + [s+1]\sin\theta)/([s+1]^2+w^2)$
10. $\dfrac{2ws^2}{(s^2+w^2)^2}$
11. $\dfrac{1}{s(s+1)^2}$
12. $\dfrac{k}{s^2+4k^2}$
13. $\dfrac{2s(s^2-3w^2)}{(s^2+w^2)^3}$
14. $\dfrac{s}{(s^2+a^2)^2}$
15. $\dfrac{6s^2-29s+36}{(s-2)^3}$

16. $\frac{s^2-a^2}{(s^2+a^2)^2}$
17. $\frac{2}{(s-3)^3}$
18. $\frac{3s-24}{s^2+4s+40}$
19. $\frac{16}{s^4-16}$
20. $\frac{s+4}{s^2+8s+12}$
21. $\frac{2}{(s+1)(s^2+2s+5)}$
22. $\frac{s^2-32}{s(s^2-64)}$
23. $\frac{s^2-2s+4}{s(s^2+4)}$
24. $\frac{2s^3-6s}{(s^2+1)^3}$
25. $\frac{6s^4-36s^2+6}{(s^2+1)^4}$
26. $\frac{4s}{(s^2-4)^2}$
27. $\frac{1}{2}\ln\left\{\frac{s+1}{s-1}\right\}$
28. $\ln\left\{\frac{s+3}{s+2}\right\}$
29. $\frac{1}{2}\ln\left\{\frac{s^2+b^2}{s^2+a^2}\right\}$
30. $\frac{2w(s+2)}{(s^2+4s+4+w^2)^2}$

Test Exercise II (Page 66)

1. (i) $\frac{1}{2}.\,e^{5t/2}$

   (ii) $3\cos 3t-\frac{10}{3}\sin 3t$

   (iii) $2t^2.\,e^{3t}$

   (iv) $5\cosh 2t-2\sinh 2t$

2. (i) $2e^{t}+e^{2t}-3e^{-3t}$

   (ii) $\frac{1}{6}\left\{e^{3t}-4+3\cos t-3\sin t\right\}$

   (iii) $\frac{1}{9}\left\{2e^{t}-2e^{-2t}+3t.\,e^{-2t}\right\}$

   (iv) $e^{-t}\left\{1-2t+\frac{t^2}{2}\right\}$

3. (i) $e^{-4t}(1-t)$

   (ii) $e^{2t}\left\{3\cos 4t+\sin 4t\right\}$

   (iii) $\frac{1}{2}\left\{21\,e^{8t}-13\,e^{4t}\right\}$

   (iv) $4\,e^{3t}-e^{-t}$

Further Problems II (Page 67)

1. $e^{-2t}\left\{\cos 2t-\sin 2t\right\}$
2. $e^{3t}(2t+7)$
3. $-2e^{-4t}+e^{-t}(2\cos 2t-\frac{1}{2}\sin 2t)$
4. $2e^{-3t}(1-t)$
5. $e^{3t}\left(\cos 2t+\frac{3}{2}\sin 2t\right)$
6. $2\,e^{-at}-1$
7. $\frac{1}{8}t\sin 4t$
8. $\frac{1}{8}\left\{e^{2t}-1-2t-2t^2\right\}$
9. $\frac{1}{b}e^{-at}\left\{b\cos bt-a\text{ siin }bt\right\}$

10. $-4\,e^{-3t} + 4\,e^{-t}\cos t - 3\,e^{-t}\sin t$
11. $1 - 2\sinh 5t$
12. $3\,e^{-2t}\cos(t\sqrt{7}) + \dfrac{5}{\sqrt{7}}\,e^{-2t}\sin(t\sqrt{7})$
13. $\dfrac{1}{39}\left\{3 - 2\,e^{-2t}\sin 3t - 3\,e^{-2t}\cos 3t\right\}$
14. $e^{-3t}(\cos t + 4\sin t)$
15. $e^{-2t}(2\cos 5t - \sin 5t)$
16. $e^{-2t}\,t\sin t$
17. $e^{t}(\cos t + \sin t)$
18. $\dfrac{1}{10}\left(23\,e^{7t} + 7\,e^{-3t}\right)$
19. $\dfrac{e^{-t/2}}{\sqrt{3}}\left\{\sqrt{3}\cos\dfrac{\sqrt{3}.t}{2} + \sin\dfrac{\sqrt{3}.t}{2}\right\}$
20. $2\cosh 3t - \dfrac{5}{3}\sinh 3t$
21. $\frac{1}{5}e^{-t}\left(4\cos t - 3\sin t\right) - \dfrac{4}{5}\,e^{-3t}$
22. $2\,e^{3t} + \cos t + \sin t$
23. $\dfrac{3}{4}\cos\dfrac{5t}{2} - \dfrac{4}{5}\sin\dfrac{5t}{2}$
24. $\dfrac{3}{50}\,e^{3t} - \dfrac{1}{25}\,e^{-2t} - \dfrac{1}{50}\,e^{-t}\cos 2t + \dfrac{9}{25}\,e^{-t}\sin 2t$
25. $\dfrac{t}{4} - \dfrac{1}{3}\sin t + \dfrac{1}{24}\sin 2t$
26. $1 + t - \cos t - \sin t$
27. $\dfrac{1}{8}t\,e^{-3t}\sin 4t$
28. $-\sin t + \sqrt{2}\sin(t\sqrt{2})$
29. $\dfrac{2t}{3} + \dfrac{1}{9} - \dfrac{1}{9}e^{-3t}$
30. $e^{4t} + e^{-t}(\cos t + 2\sin t)$

Test Exercise III (Page 102)

1. $x = e^{-2t}$
2. $x = \dfrac{1}{73}\left\{\dfrac{105}{4}e^{3t/4} + 8\cos 2t - 3\sin 2t\right\}$
3. $x = 5(e^{3t} - e^{2t})$
4. $x = \dfrac{2}{5} + \dfrac{5}{4}e^{-t} - \dfrac{3}{4}e^{-5t}$
5. $x = \dfrac{29}{16}e^{2t} + \dfrac{5}{16}e^{-2t} - \dfrac{1}{8}\cos 2t$
6. $x = e^{t}\left(1 - t + \dfrac{t^3}{6}\right)$
7. $x = \dfrac{e^{2t}}{2}\left\{3\sin t - \cos t\right\} + \dfrac{1}{2}e^{3t}$
8. $x = \dfrac{1}{5}e^{2t} - \dfrac{1}{2}e^{t} + \dfrac{1}{10}\left\{3\cos t + 11\sin t\right\}$

Further Problems III (Page 103)

1. $x = \dfrac{6}{5}e^{2t} - \dfrac{3}{2}e^{t} + \dfrac{3}{10}\cos t + \dfrac{1}{10}\sin t$
2. $x = \cos 2t - \dfrac{1}{8}\sin 2t + \dfrac{t}{4}$
3. $x = 4 + 5\,e^{-t} - 2\,e^{-3t}$
4. $y = \dfrac{e^{-2t}}{20}\left\{39\cos 2t + 47\sin 2t\right\} + \dfrac{1}{10}\sin 2t + \dfrac{1}{20}\cos 2t$
5. $x = e^{t}\left\{\dfrac{t^2}{2} - t - 2\right\}$
6. $y = \dfrac{7}{2}e^{-t} + \dfrac{9}{2}t\,e^{-t} - \dfrac{1}{2}\cos t$
7. $x = 2t^2 - 6t + 7 - 8\,e^{-t} + e^{-2t}$
8. $x = e^{-2t}(9\sin t + 5\cos t) + \sin t - \cos t$
9. $y = \left\{\dfrac{1}{8} + \dfrac{3t}{4}\right\}e^{-2t} + \dfrac{t^2\,e^{-2t}}{2} + \dfrac{3}{8} - \dfrac{t}{2} + \dfrac{t^2}{4}$

10. $x = -\frac{e^{-t}}{4} + \frac{1}{4}\cos t + \frac{t}{4}(\sin t - \cos t)$

11. $y = e^{3t}\left(A + \frac{t}{12}\right) + e^{-3t}\left(B - \frac{t}{12}\right) - \frac{1}{81}(9t^2 + 2)$

12. $y = Ae^{3t} + Be^{-2t} + \frac{e^{3t}}{250}(25t^2 - 10t + 2)$

13. $y = e^{-t}\left(A + \frac{1}{2}\sin t - \frac{1}{2}\cos t\right) + B\,e^{-2t}$

14. $y = e^t + e^{2t}(1 + 6t - 3t^2 + t^3 + \sin t - \cos t)$

15. $x = \frac{a\,e^{-kt}}{n}(n \cos nt + k \sin nt)$

17. $F(t) = e^{2t} - 7\,e^{-t} + 6\,e^{-2t}(\cos t + \sin t)$

18. $y = \frac{4}{5}(e^{-4t} - 1)\cos 4t + \frac{2}{5}(e^{-4t} + 1)\sin 4t$

20. $x = e^{-kt}\left(A \cos nt + \left[B + \frac{t}{2n}\right]\sin nt\right)$

Test Exercise IV (Page 131)

1. $x = \frac{1}{5}\left\{e^t + 2\,e^{-t} - 3\cos 2t + 3\sin 2t\right\}$

   $y = \frac{1}{5}\left\{3\sin 2t + 3\cos 2t - 2\,e^t - e^{-t}\right\}$

2. $x = \frac{3\,e^{2t}}{2}\left(2 + t\right)$

   $y = e^{2t}(1 + t)$

3. $x = A\,e^t + B\,e^{4t}$

   $y = A\,e^t - 2\,B\,e^{4t}$

4. $x = \frac{3}{2}\left\{e^{-t} - e^t + 2t\,e^t\right\}$

5. $y = \frac{2}{3}\left(\cos t - \cos 2t\right)$

Further Problems IV (Page 132)

1. $x = \left\{-\frac{t}{3} + \frac{19}{36}\right\}e^{-4t} - \frac{e^{-2t}}{4} + \frac{2\,e^{-t}}{9}$

   $y = \left\{\frac{t}{3} - \frac{7}{36}\right\}e^{-4t} + \frac{3\,e^{-2t}}{4} + \frac{e^{-t}}{9}$

2. $x = -\frac{1}{3}\left\{3 + e^{-4t} - 14e^{-t} - 10te^{-t}\right\}$

   $y = \frac{1}{3}\left\{9 - 20te^{-t} - e^{-4t} - 8e^{-t}\right\}$

3. $x = 2 + \frac{t^2}{2} + \frac{e^{-t}}{2} - \frac{3}{2}\sin t + \frac{1}{2}\cos t$

   $y = 1 - \frac{e^{-t}}{2} + \frac{3}{2}\sin t - \frac{1}{2}\cos t$

4. $x = \frac{1}{6}\sin 4t - \sin t + \frac{2}{3}\sin 2t$

   $y = -\frac{1}{6}\cos 4t + \cos t - \frac{5}{6}\cos 2t$

5. $x = \frac{1}{4}\left\{3e^t + 7e^{-t}\right\} - \frac{1}{10}\left\{19\cos t - 2\sin t\right\} + \frac{2}{5}e^{2t}$

   $y = -\frac{1}{12}\left\{3e^t + 7e^{-t}\right\} + \frac{1}{10}\left\{19\cos t - 2\sin t\right\} - \frac{1}{15}e^{2t}$

6. $x = 3\sin t - 2\cos t + e^{-2t}$

   $y = -\frac{7}{2}\sin t + \frac{9}{2}\cos t - \frac{1}{2}e^{-3t}$

7. $x = -\frac{11}{16}e^t + \frac{15}{16}e^{3t} - \frac{1}{4}e^{-2t}$

   $y = -\frac{1}{8}e^t + \frac{3}{8}e^{3t}$

8. $x = \dfrac{E}{R}\left\{\dfrac{2}{3} - \dfrac{1}{2}e^{-Rt/L} - \dfrac{1}{6}e^{-3Rt/L}\right\}$

   $y = \dfrac{E}{R}\left\{\dfrac{1}{3} - \dfrac{1}{2}e^{-Rt/L} + \dfrac{1}{6}e^{-3Rt/L}\right\}$

9. $x = 5e^{t} + 3e^{-t}$

   $y = 4e^{t} - e^{-t}$

10. $x = 2\ \{e^{2t} - e^{t}\}$

    $y = 5e^{2t} - 4e^{t}$

11. $x = e^{-t} + e^{-2t} - 2e^{-3t/2}$

    $y = -\ 2e^{-t} - e^{-2t} + 3e^{-3t/2}$

12. $x = \dfrac{1}{9}e^{-t} - \dfrac{1}{9}e^{2t} + \dfrac{1}{3}te^{-t}$

    $y = \dfrac{1}{9}e^{-t} + \dfrac{4}{45}e^{2t}$

    $\quad -\dfrac{1}{5}\cos t - \dfrac{2}{5}\sin t + \dfrac{1}{3}t\,e^{-t}$

14. $x = -\dfrac{5}{3}\{\cosh\sqrt{2}t + \sqrt{2}\sinh\sqrt{2}t\}$

    $\quad + \dfrac{5}{3}\ \{\cos 2t + \sin 2t\}$

15. $\theta = a\cos\left(t\sqrt{\dfrac{3g}{10a}}\right) + \dfrac{3a}{4}\cos\left(t\sqrt{\dfrac{6g}{a}}\right)$

    $\phi = \dfrac{5a}{4}\cos\left(t\sqrt{\dfrac{3g}{10a}}\right) - \dfrac{a}{4}\cos\left(t\sqrt{\dfrac{6g}{a}}\right)$

16. $x = \dfrac{umc}{He}\sin\left(\dfrac{He}{mc}t\right)$

    $y = -\dfrac{umc}{He}\left\{1 - \cos\left(\dfrac{He}{mc}t\right)\right\}$

17. $y = \dfrac{3}{5}\sin 2t + \dfrac{4}{15}\sin 3t$

    $\quad -\dfrac{4}{5}\cos 2t + \dfrac{48}{35}\cos 3t - \dfrac{4}{7}\cos 4t$

18. $x = \dfrac{3a}{4}\cos t + \dfrac{a}{4}\cos\sqrt{3}\,.\,t$

    $y = \dfrac{3a}{4}\cos t - \dfrac{a}{4}\cos\sqrt{3}\,.\,t$

19. $x = \dfrac{a}{2}(\sin t - \sinh t)$

    $y = \dfrac{a}{2}(\cosh t - \cos t)$

20. $i_1 = 1 + e^{-35t/2}$

    $i_2 = 2\ \{1 + e^{-35t/2}\}$

Test Exercise V (Page 170)

1.(i)

$\mathcal{L}\ \{F(t)\} = \dfrac{2}{s^2}\ (1 - e^{-2s})$

(iii)

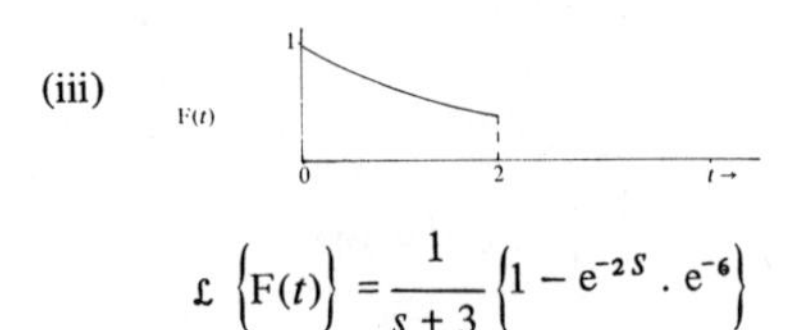

$\mathcal{L}\ \{F(t)\} = \dfrac{1}{s+3}\{1 - e^{-2s}\ .\ e^{-6}\}$

(ii)

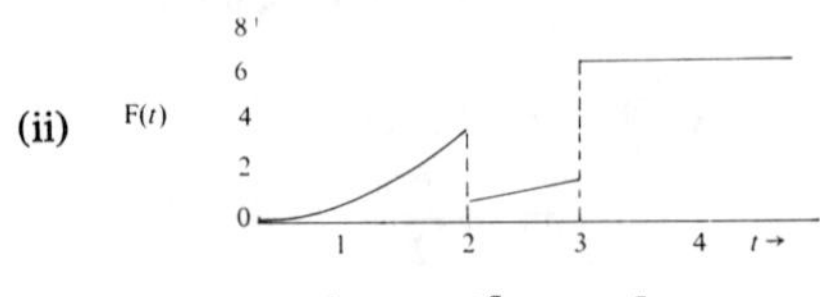

$\mathcal{L}\ \{F(t)\} = \dfrac{2}{s^3} - \dfrac{2e^{-2s}}{s^3} - \dfrac{3e^{-2s}}{s^2} - \dfrac{3e^{-2s}}{s}$

$\quad - \dfrac{e^{-3s}}{s^2} + \dfrac{5e^{-3s}}{s}$

(iv)

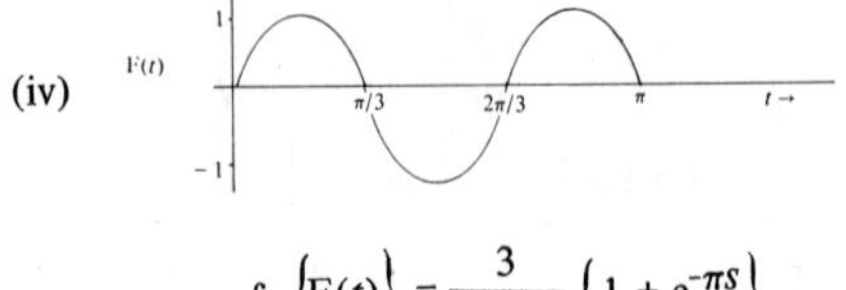

$\mathcal{L}\ \{F(t)\} = \dfrac{3}{s^2+9}\{1 + e^{-\pi s}\}$

2. $F(t) = 3 \,.\, H(t) - 1 \,.\, H(t-1) + 6 \,.\, H(t-2)$

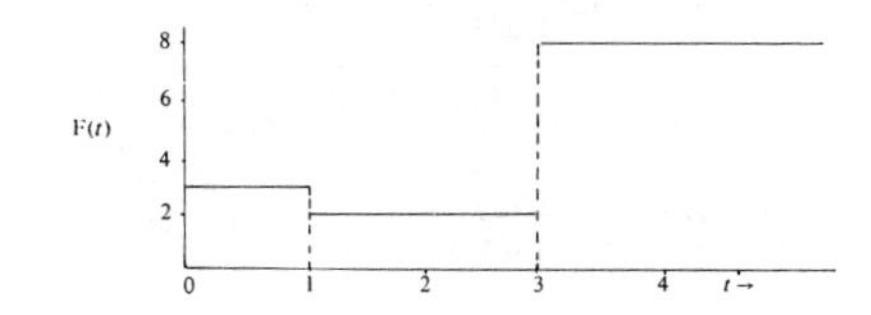

3. $F(1) = 1; \qquad F(4) = 4; \qquad F(6) = 6$

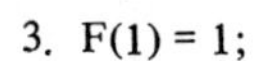

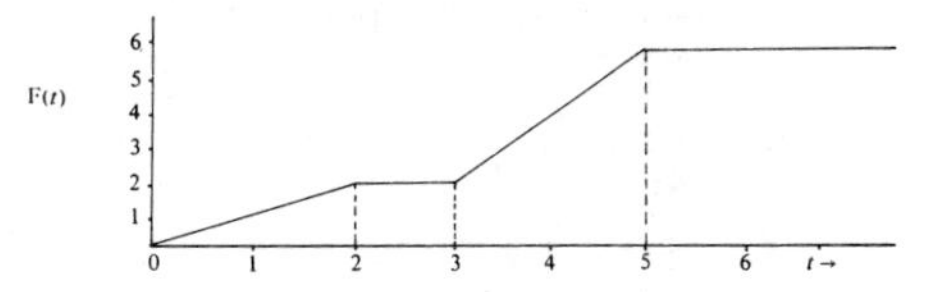

Further Problems V (Page 171)

1. (a) $\dfrac{e^{-3s}}{2}\left\{\dfrac{e^{6}}{s-2} - \dfrac{e^{-6}}{s+2}\right\}$

   (b) $-\dfrac{s}{s^2+1}\,e^{-\pi s}$

   (c) $\dfrac{2e^{-2s}}{s^3}\left\{1 + 2s + 2s^2\right\}$

2. (a) $F(t) = \cos t \,.\, H(t) - \cos t \,.\, H(t-\pi)$

   $+ \cos t \,.\, H(t-2\pi) \ldots$

   (b) $F(t) = \dfrac{1}{2} e^{3(t-2)}$

   $\sin 2(t-2) \,.\, H(t-2)$

   (c) $F(t) = \dfrac{1}{2}(t-2)^2$

   $e^{-3(t-2)} \,.\, H(t-2)$

   (d) $F(t) = H(t) + H(t-\pi)$

   $+ H(t-2\pi) + \ldots$

   (e) $F(t) = \dfrac{1}{2}(t-3)^2$

   $e^{4(t-3)} \,.\, H(t-3)$

3. $\mathcal{L}\left\{F(t)\right\} = \dfrac{1}{s(1-e^{-\pi s})}$

4. $F(t) = 4 \,.\, H(t) - 5(t-1) \,.\, H(t-1)$

   $+ 5(t-3) \,.\, H(t-3)$

5. $i = \left\{5e^2 e^{-t} + e^{-10}\, e^{5t} - 6\right\} \,.\, H(t-2)$

6. $\dfrac{s + (s-1)\,e^{-\pi s}}{s^2+1}$

7. $\dfrac{2(1-e^{-\pi s})}{s^2+4}$

8. (a) $F(t) = t^2 \,.\, H(t) - (t^2 - 4t) \,.\, H(t-2)$

   (b) $F(t) = \sin t \,.\, H(t)$

   $+ (\sin 2t - \sin t) \,.\, H(t-\pi)$

   $+ (\sin 3t - \sin 2t) \,.\, H(t-2\pi).$

9. $\left\{\dfrac{1}{s^2} + \dfrac{1}{s}\right\} e^{-s} - \left\{\dfrac{1}{s^2} + \dfrac{2}{s}\right\} e^{-2s}$

10. $\dfrac{2}{s^3} - e^{-2s}\left\{\dfrac{3}{s} + \dfrac{3}{s^2} + \dfrac{2}{s^3}\right\} + e^{-3s}\left\{\dfrac{5}{s} - \dfrac{1}{s^2}\right\}$

11. $y = \displaystyle\sum_{n=0}^{\infty} H(t-n\pi) -$

    $\displaystyle\sum_{n=0}^{\infty} \cos(t-n\pi) \,.\, H(t-n\pi)$

13. $x = 3 - 2\cos t$

    $+ 2\left\{t - 4 - \sin(t-4)\right\}. \quad H(t-4)$

Test Exercise VI (Page 190)

1. (a) $\dfrac{1}{s(1-e^{-4s})}\left\{2-3e^{-2s}+e^{-4s}\right\}$

   (b) $\dfrac{5}{4s^2(1-e^{-8s})}\left\{1-e^{-4s}-4s\,e^{-4s}\right\}$

   (c) $\dfrac{4(1+e^{-\pi s})}{(s^2+1)(1-e^{-\pi s})}$

2. F(t): 2, 0, 5, 10, t→

   $\mathcal{L}\left\{F(t)\right\} = \dfrac{2}{(s+1)(1-e^{-5s})}\left\{1-e^{-5}.e^{-5s}\right\}$

3. $F(t) = 2\,.\,H(t) + 4\,.\,H(t-2) - 4\,.\,H(t-3) + 4\,.\,H(t-5) - \ldots$

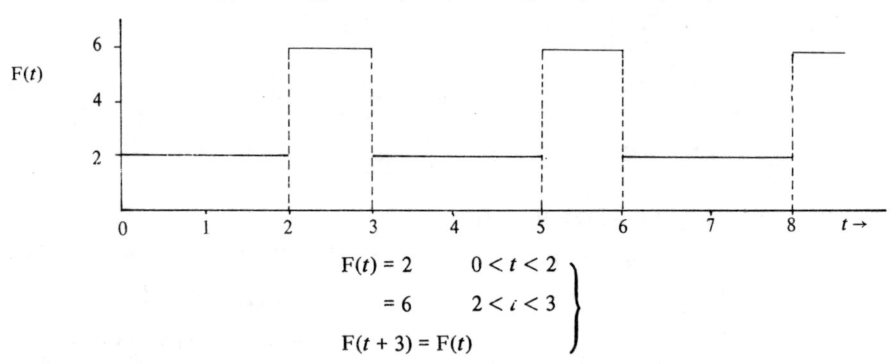

$$\left.\begin{aligned} F(t) &= 2 \qquad 0<t<2\\ &= 6 \qquad 2<t<3\\ F(t+3) &= F(t)\end{aligned}\right\}$$

Further Problems VI (Page 191)

1. $\dfrac{1}{1-e^{-3s}}\left\{\dfrac{2}{s^3}-\dfrac{2\,e^{-2s}}{s^3}-\dfrac{4\,e^{-2s}}{s^2}-\dfrac{4\,e^{-3s}}{s}\right\}$

2. $\dfrac{1}{1-e^{-2s}}\left\{\dfrac{1-e^{-(s-2)}}{s-2}\right\}+\dfrac{e^{-s}}{s(1+e^{-s})}$

3. $\dfrac{1-e^{-s}(s+1)}{s^2(1-e^{-2s})}$

4. $\dfrac{1}{s^2+1}\cdot\dfrac{1}{1-e^{-\pi s}}$

5. $\dfrac{1}{s^2}-\dfrac{2\pi e^{-2\pi s}}{s(1-e^{-2\pi s})}$

6. $\dfrac{E}{s}\tanh\left\{\dfrac{\pi s}{2w}\right\}$

7. $\dfrac{1}{s}\tanh\left\{\dfrac{as}{2}\right\}$

8. $\dfrac{1}{s^2}-\dfrac{w}{s}\left\{\dfrac{e^{-ws}}{1-e^{-ws}}\right\}$

9. $\dfrac{1-e^{-\pi s}-\pi s\,e^{-\pi s}}{s^2(1-e^{-2\pi s})}$

10. $\dfrac{e^{2(1-s)\pi}-1}{(1-s)(1-e^{-2\pi s})}$

11. $\dfrac{1}{as^2}-\dfrac{e^{-as}}{s(1-e^{-as})}$

12. $\dfrac{2(e^{2\pi s}-1-2\pi s-2\pi^2 s^2)}{s^3(e^{2\pi s}-1)}$

13. $\dfrac{2\pi a T\coth\left(\dfrac{Ts}{4}\right)}{(s^2T^2+4\pi^2)}$

14. $y = \dfrac{c}{w^2}\ \cos\dfrac{wT}{2} - \cos wT)$

15. $y = e^{-t} - e^{-2t}$

    $+\left(1-\dfrac{1}{e}\right)\left\{(e+e^2)e^{-t}-(e^2+e^4)e^{-2t}\right\}$

Test Exercise VII (Page 209)

1. (i) e^{-8} (ii) -2 (iii) 31

2. (i) $5\,e^{-2s}$ (ii) $e^{-\pi s/2}$ (iii) $e^{-(s+2)}$

3.

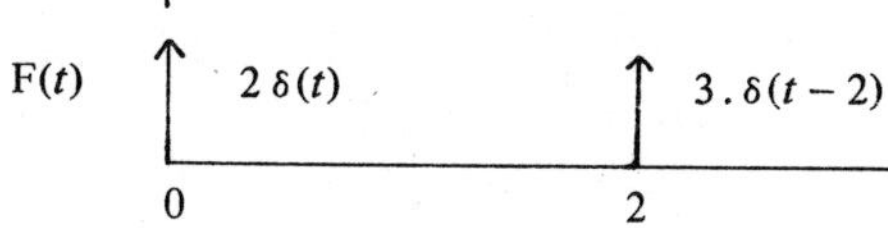

5 $t \rightarrow$

$4 \,.\, \delta(t-5)$

$\mathcal{L}\,\{F(t)\} = 2 + 3e^{-2s} - 4e^{-5s}$

4. $x = e^{-t}\,(3\cos 2t + 4\sin 2t)$

5. $x = e^{-t} - e^{-3t} + e^{6}\,.\,e^{-t}\,.\,H(t-6)$
$\qquad - e^{18}\,.\,e^{-3t}\,.\,H(t-6)$

Further Problems VII (Page 210)

1. $e^{-s}(s^2 + s + 1)/s^2$

2. $x = 4\,e^{-2t} + 3\,e^{-3t}$
$y = -\,e^{-2t} - e^{-3t}$

3. $x = \dfrac{P}{Mw}\sin wt$

4. $i = \dfrac{E}{L}\cos\dfrac{t}{\sqrt{(LC)}}$

5. $i = \dfrac{E}{La}\,e^{-Rt/2L}\left\{a\cos at - \dfrac{R}{2L}\sin at\right\}$

where $a^2 = \dfrac{1}{LC} - \dfrac{R^2}{4L^2}$

6. $x = \dfrac{I}{mw}\,\{\sin wt\,.\,H(t)$
$\qquad - \cos wt\,.\,H(t - \pi/2w)\}$

8. $v = \dfrac{E}{wLC}\,e^{-Rt/2L}\sinh wt$

where $w^2 = \dfrac{R^2}{4L^2} - \dfrac{1}{LC}$.

Test Exercise VIII (Page 239)

1. $i = \dfrac{1}{9}\left\{e^{-2t} - \cos\sqrt{5}\,.\,t + \sqrt{2}\sin\sqrt{5}\,.\,t\right\}$

2. $F(t) = -\dfrac{1}{20} + \dfrac{5}{36}e^{4t} - \dfrac{4}{45}e^{-5t}$

3. $F(t) = 1 - 2t - 2t^2 - t^3/6$

4. (i) 0 (ii) $\dfrac{E}{R}$

5. $t\sin 2t$

Further Problems VIII (Page 240)

1. (a) $\frac{1}{6} - \frac{1}{2} e^{-t} + \frac{1}{2} e^{-2t} - \frac{1}{6} e^{-3t}$

   (b) $-\frac{1}{6} e^{-t} - \frac{4}{3} e^{2t} + \frac{7}{2} e^{3t}$

   (c) $2\, e^{t} - 2 \cos t + \sin t$

   (d) $\frac{1}{5} e^{-t} \{4 \cos t - 3 \sin t\} - \frac{4}{5} e^{-3t}$

   (e) $\frac{1}{50} \{3\, e^{3t} - 2\, e^{-2t} - e^{-t} \cos 2t$
   $+ 18\, e^{-t} \sin 2t\}$

2. (a) $\frac{1}{16} \{e^{2t} - e^{-2t} - 4te^{-2t}\}$

   (b) $\frac{1}{2} e^{t} - \frac{1}{2} \cos t + \frac{1}{2} \sin t$

   (c) $t^2 + 2 \cos t - 2$

   (d) $\frac{1}{4} \left\{ e^{-2t} + 2t - 1 \right\}$

   (e) $\frac{1}{2} t \sin t$

   (f) $te^{-t} + 2\, e^{-t} + t - 2$

   (g) $\frac{1}{8} \{(3 - t^2) \sin t - 3t \cos t\}$

INDEX